Renewable Energy

Edited by Roland Wengenmayr
and Thomas Bührke

Related Titles

Würfel, P.

Physics of Solar Cells

From Basic Principles to Advanced Concepts

2009

ISBN: 978-3-527-40857-3

Abou-Ras, D., Kirchartz, T., Rau, U. (Hrsg.)

Advanced Characterization Techniques for Thin Film Solar Cells

2011

ISBN: 978-3-527-41003-3

Scheer, R., Schock, H.-W.

Chalcogenide Photovoltaics

Physics, Technologies, and Thin Film Devices

2011

ISBN: 978-3-527-31459-1

Stolten, D. (Hrsg.)

Hydrogen and Fuel Cells

Fundamentals, Technologies and Applications

2010

ISBN: 978-3-527-32711-9

Vogel, W., Kalb, H.

Large-Scale Solar Thermal Power

Technologies, Costs and Development

2010

ISBN: 978-3-527-40515-2

Huenges, E. (Hrsg.)

Geothermal Energy Systems

Exploration, Development, and Utilization

2010

ISBN: 978-3-527-40831-3

Keyhani, A., Marwali, M. N., Dai, M.

Integration of Green and Renewable Energy in Electric Power Systems

2010

ISBN: 978-0-470-18776-0

Olah, G. A., Goeppert, A., Prakash, G. K. S.

Beyond Oil and Gas: The Methanol Economy

2010

ISBN: 978-3-527-32422-4

Renewable Energy

Sustainable Concepts for the Energy Change

Edited by
Roland Wengenmayr and Thomas Bührke

2nd Edition

**WILEY-
VCH**

WILEY-VCH Verlag GmbH & Co. KGaA

The Editors

Roland Wengenmayr
Frankfurt/Main, Germany

Thomas Bührke
Schwetzingen, Germany

**German edition and additional articles
translated by:**
Prof. William Brewer

Library of Congress Card No.:
applied for

British Library Cataloguing-in-Publication Data
A catalogue record for this book is available from the
British Library.

**Bibliographic information published by
the Deutsche Nationalbibliothek**
Die Deutsche Nationalbibliothek lists this publication in
the Deutsche Nationalbibliografie; detailed bibliographic
data are available in the Internet at
<http://dnb.d-nb.de>.

Typesetting   TypoDesign Hecker GmbH, Leimen

Printing and Binding   Himmer AG, Augsburg

Cover Design   Bluesea Design, Simone Benjamin,
McLeese Lake, Canada

Printed in the Federal Republic of Germany
Printed on acid-free paper

ISBN   978-3-527-41187-0

Foreword

Today, it is generally recognized that human activities are significantly changing the composition of the earth's atmosphere and are thus provoking the imminent threat of catastrophic climate change. Critical concentration changes are those of carbon dioxide (CO_2), laughing gas (dinitrogen monoxide, N_2O) and methane (CH_4). The present-day concentration of CO_2 is above 380 ppm (parts per million), far more than the maximum CO_2 concentration of about 290 ppm observed for the last 800,000 years. The most recent reports of the World Climate Council, the Intergovernmental Panel on Climate Change (IPCC) and the COP-16 meeting in Cancun in December, 2010 demonstrate that the world is beginning to face the technological and political challenges posed by the requirement to reduce the emissions of these gases by 80 % within the next few decades. The nuclear power plant catastrophe in Fukushima on March 11th, 2011 showed in a drastic way that nuclear power is not the correct path to CO_2-free power production. Germany made a reversal of policy as a result, which has attracted attention worldwide. In the coming years, we shall certainly be trailblazers in the global transformation of our energy system in the direction of one hundred percent renewable sources.

This ambitious goal can be achieved only through substantial progress in the two main areas that affect this issue: Rapid growth of energy production from renewable sources, and increased energy efficiency, especially of buildings which cause a large portion of our total energy needs. Unfortunately, these two concrete, positive goals are still being neglected in the international climate negotiations.

This book presents a comprehensive treatment of these critical objectives. The 26 chapters of this greatly expanded 2nd English edition have been written by experts in their respective fields, covering the most important issues and technologies needed to reach these dual goals. This volume provides an excellent, concise overview of this important area for interested general readers, combined with interesting details on each topic for the specialists.

The topics addressed include photovoltaics, solar-thermal energy, geothermal energy, energy from wind, waves, tides, osmosis, conventional hydroelectric power, biogenic energy, hydrogen technology with fuel cells, building efficiency and solar cooling. The very topical question of how automobile mobility can be combined with sustainable energies is discussed in a chapter on electric vehicles. The treatment of biogenic energy sources has been expanded in additional chapters.

In each chapter, the detailed discussion and references to the current literature enable the reader to form his or her own opinion concerning the feasibility and potential of these various technologies. The volume appears to be well suited for generally interested readers, but may also be used profitably in advanced graduate classes on renewable energy. It seems especially well suited to assist students who are in the process of selecting an inspiring, relevant topic for their studies and later for their thesis research.

Eicke R. Weber,
Director,
ISE Institute for Solar Energy Systems,
Freiburg, Germany

Preface

This book gives a comprehensive overview of the development of renewable energy sources, which are essential for substituting fossil fuels and nuclear energy, and thus in securing a healthy future for our earth.

A variety of energy resources have been discussed by experts from each of the fields to provide the readers with an insight into the state of the art of sustainable energies and their economic potential.

Most important is that:

1) Some of the renewable energy sources are already less expensive than oil or nuclear power in their overall economic balance today, such as wind power or solar thermal energy; close to achieving this goal, for example, are also solar cell panels.

2) It is misleading to seek an attractive alternative in nuclear power plants: They are not! By comparison, the construction of a wind park takes under one year, while the construction of a nuclear power plant requires close to seven years. The cost of a wind park is less than 30 % of the price of a nuclear power plant of the same output. The nuclear plant also entails additional costs for later dismantling and for the final storage of its radioactive waste products, which will put a burden on our descendants for many hundreds of years to come. It is also a little-known fact that the uranium mines – most of them in the Third World – contaminate large areas with their radioactive wastes and poison rivers with millions of tons of toxic sludge.

The good news is that already at the end of 2010, worldwide annual power generation by wind plants and solar cells exceeded the output of all the nuclear power plants in the USA and France combined. However, representatives of the conventional power industry frequently argue that solar conversion is unreliable because the sun doesn't always shine. We give them an emphatic answer: "No, at night of course not, but who needs more energy at night when there is already more low-cost power available than we can use?" An important point is that solar cells – seen from a worldwide perspective – can make an essential contribution in the midday and afternoon periods, when power consumption is highest. Wind, however, fluctuates more, but with more digitally-regulated power distribution and rapidly developing storage facilities, these fluctuations can be minimized, and already today, wind parks are important contributors to the global power balance.

Today, the nuclear and petroleum industries take the growing competition from wind and solar energy very seriously. In the USA, one is made aware of this by their alarmingly accelerating lobbyist activities in Washington, opposing support of the development of sustainable energies. We in the democratic countries should use our voting power to elect those parties and politicians who understand the necessities of our times and thus the opportunities of sustainable energies, and who support and work towards their further development and implementation.

This book offers a good choice of topics to all its interested readers who want to inform themselves more thoroughly, and in addition to all those who want to work in one of the many branches of sustainable energy development and deployment. It represents an important contribution towards advancing their urgently needed implementation and thereby avoiding a threatening catastrophe brought on by unwise energy policy.

All together, it is a pleasure to read this book; it deserves a special place on every bookshelf, with its excellent form and content. It will have a lasting value in recording the current state of the rapid developments of sustainable energies.

Karl W. Böer,
Distinguished Professor of Physics and Solar Energy, emeritus
University of Delaware

First-hand Information

In the four years since the publication of the first edition of this book, the world has undergone drastic changes in terms of energy. This is reflected in the expansion of this second edition to nearly 30 chapters. The most dramatic occurrence was the terrible Tsunami which struck Japan in March of 2011 and set off a reactor catastrophe at the nuclear power plants in Fukushima. In Germany, the government reacted by deciding to phase out nuclear power completely by 2022. Nevertheless, the ambitious German goals for reducing the emissions of greenhouse gases were retained. Renewable energy sources will therefore have to play an increasing role in the coming years.

Nearly four hundred thousand jobs have been created in Germany in the field of sustainable energy, many of them in the area of wind energy. However, the German photovoltaic industry is in crisis, in part because Chinese solar-module producers can now manufacture and market their products at a lower price. In 2012, the U. S. Deparment of Commerce posted anti-dumping duties on solar cells from China. This conflict illustrates what basically is good news for the world as a whole, since the increased competition will rapidly lower the costs of solar power.

This book of course is not restricted to only the German perspective. In particular, it introduces a variety of technologies which can help the world to make use of sustainable energies. From a technical point of view, this field is extremely dynamic. This can be seen by again looking at the example of photovoltaic power: Since the first edition, the established technologies based on silicon have encountered increasing competition from thin-film module manufacturers, whose products save on energy and resources. Accordingly, Nikolaus Meyer completely revised his chapter on chalcopyrite (CIS) solar cells. The chapter by Michael Harr, Dieter Bonnet and Karl-Heinz Fischer on the promising cadmium telluride (CdTe) thin-film solar cells is completely new in this edition.

The biofuels industry, on the other hand, has developed an image problem. Aside from the competition for arable land with food-producing agriculture (the 'food or fuel' controversy), the first generation of biofuels has also been pilloried because of its poor CO_2 balance. Gerhard Kreysa gives an extensive analysis of the contribution that can be made by biofuels to the world's energy supply in a reasonable and sustainable way. Nicolaus Dahmen and his collaborators introduce their environmentally friendly bioliq® process from the Karlsruhe Institute of Technology, which is on the threshold of commercialization and has aroused interest internationally. Carola Griehl's research team looks forward to a future powered by biofuels produced from algae.

Electric power from renewable sources requires intelligent distribution and storage. An exciting international example is the DESERTEC project, which envisions a supply of power to Europe from the sunny regions of North Africa. Franz Trieb from the German Aeronautics and Space Research Center was involved in the DESERTEC feasibility study and presents its results in detail here, in particular the win-win situation for both producers and consumers. The large solar thermal plants can meet the rapidly growing power needs of the North African population, for example for supplying potable water by desalination of seawater.

Nearly all the chapters were written by professionals in the respective fields. That makes this book an especially valuable and reliable source of information. It can be readily understood by those with a general educational background. Only a very few chapters include a small amount of mathematics. We have left these formulas intentionally for those readers who want to delve more deeply into the material; these few short passages can be skipped over without losing the thread of information. Extensive reference lists and web links (updated shortly before printing) offer numerous opportunities to access further material on these topics.

All the numbers and facts have been carefully checked, which is not to be taken for granted. Unfortunately, there is much misinformation and misleading folklore in circulation regarding sustainable energies. This book is therefore intended to provide a reliable and solid source of information, so that it can also be used as a reference work. Its readers will be able to enter into informed discussions and make competent decisions about these important topics.

We thank all of the authors for their excellent cooperation, William Brewer for his careful translation, and the publishers for this beautifully designed and colorful book. In particular, we want to express our heartfelt thanks to Ulrike Fuchs of Wiley-VCH Berlin for her active support and her patience with us. Without her, this wonderful book would never have been completed.

Thomas Bührke and Roland Wengenmayr

Schwetzingen and Frankfurt am Main, Germany
August 2012

Contents

Photo: DLR

Photo: Voith Hyd

CONTENTS

Photo: Vestas Central Europe

Poto: GFZ

This large photovoltaic roof installation, above the Munich Fairgrounds building, has a nominal power output of about 1 MWel. *(Photo: Shell Solar).*

Renewable Energy Sources – a Survey

BY HARALD KOHL | WOLFHART DÜRRSCHMIDT

Renewable energy sources have developed into a global success story. How great is their contribution at present in Germany, in the European Union and in the world? How strong is their potential for expansion? A progress report on the balance of innovation.

Renewable energy has become a success story in Germany, in Europe, the USA and Asia. Current laws, directives, data, reports, studies etc. can be found on the web site on renewable energies of the German Federal Ministry for the Environment [1].

The European Union – ambitious Goals

Let us first take a look at developments within the European Union: On June 25, 2009, Directive 2009/28/EG of the European Parliament and the Council for the Advancement of Renewable Energies in the EU took effect [2]. The binding goal of this directive is to increase the proportion of renewable energy use relative to the overall energy consumption in the EU from ca. 8.5 % in the year 2005 to 20 % by the year 2020. The fraction used in transportation is to be at least 10 % in all the member states by 2020. This includes not only biofuels, but also electric transportation using power from renewable sources. A binding goal was set for each member state for the fraction of sustainable energy in the total energy consumption (electric power, heating/cooling and transportation), depending on the starting value in that country. For Germany, this goal is 18 % by 2020, while for the neighboring countries, it is: Belgium, 13 %; Denmark, 30 %; France, 23 %; Luxemburg, 11 %; the Netherlands, 14 %; Austria, 34 %; Poland, 15 %; and for the Czech Republic, 13 %.

The member states can choose for themselves which means they employ to reach these goals. The development of renewable energy sources for electric power generation has been particularly successful in those member states

*Installation of a wind-energy plant at the offshore wind energy park **Alpha Ventus**, which started operation in the North Sea in 2009 (photo: alpha ventus).*

TAB. 1 | WORLDWIDE INSTALLED WIND POWER IN MW; YEAR 2010

Region	Country (examples)	End of 2009	New in 2010	End of 2010
Africa and	Egypt	430	120	550
Middle East	Morocco	253	33	286
total		*866*	*213*	*1 079*
Asia	China	25 805	16 500	42 287
	India	10 926	2 139	13 065
	Japan	2 085	221	2 304
total		*39 639*	*19 022*	*58 641*
Europe	Germany	25 777	1 493	27 214
(EU- and Non-	Spain	19 160	1 516	20 676
EU countries)	Italy	4 849	948	5 797
	France	4 574	1 086	5 660
	United Kingdom	4 245	962	5 204
	Austria	995	16	1 011
total		*76 300*	*9 883*	*86 075*
Latin America	Brazil	606	326	931
und Caribbean	Mexico	202	316	519
total		*1 306*	*703*	*2 008*
North America	USA	35 086	5 115	40 180
	Canada	3 319	690	4 009
total		*38 405*	*5 805*	*44 189*
Pacific	Australia	1 702	167	1 880
total		*2 221*	*176*	*2 397*
Global totals		**158 738**	**35 802**	**194 390**

Source: [10]. The values are in part preliminary due to rounding errors, shutdown of plants, and differing statistical methods, leading to deviations from national statistics.

which, like Germany, have given priority and a grid feed-in premium to power from these sources, analogous to the German Renewable Energy Act (EEG). Twenty of the EU countries have in the meantime adopted such laws to promote the use of power from renewable sources; worldwide, 50 countries have done so [3,4].

As an interim result, by 2010 the following proportions of renewable energy were used in the EU: for electricity, about 20 %; for heating/cooling, around 13 %; and for road transportation, around 4 %. Electric power generating plants, especially those using wind power, solar energy and bioenergy, have made clear progress. In the future, they will most likely maintain their head start. In this process, not only are technical progress and cost efficiency relevant, but also the establishment of organizational structures which take into account all the criteria of sustainability. Systems analysis and optimization, participation and acceptance by affected citizens, accompanying ecological research, environmental and nature protection as well as resource conservation are all becoming increasingly important. In order to reach the goals for renewable energy of 10 % of transportation and 20 % of the total energy consumption by 2020, the fraction of electric power from renewable sources must be around one-third of the total by then. A finely-meshed monitoring system was established, based on regular reports by the member states and the EU Commission [4–6].

Wind Energy is booming internationally

Especially the example of wind energy demonstrates that the rate of success can vary considerably even with comparable starting conditions. The environmental and energy-policy framework is decisive here. In particular, the German Renewable Energy Act (EEG), with its power feed-in and repayment regulations that encourage investments in renewable energy, along with similar legislation in Spain, has had considerable effect in comparison to other countries. Germany and Spain had an installed wind power of around 50,000 MW in 2010, more than half of that in the EU as a whole (with ca. 94,000 MW) [4, 7].

But not only in the EU, also in China, India and the USA, the market for wind power plants is booming (Table 1). In the past few decades, a whole new branch of engineering technology has developed. Megawatt installations are now predominant. German and Danish firms are among the leaders in this field. About three-fourths of the wind power plants manufactured in Germany are now exported. Germany has acquired a similar prominence in solar power generation, both in photovoltaics and in solar thermal technology.

Successful Energy Policies in Germany

The German example in particular shows how the efforts of individual protagonists, support via suitable instruments (research and development, Renewable Energy Act, Heat Energy Input Act, assistance for entering the market, etc.) as well as cooperation between scientific institutions and innovative industrial firms in the area of renewable energy sources

INTERNET

BMU brochure [4] and other materials
www.erneuerbare-energien.de/english

ABB. 1 | ELECTRIC POWER IN GERMANY FROM RENEWABLE ENERGY SOURCES

The time evolution of the fraction of renewable energy sources for electric power production in Germany from 1990 to 2011 (TWh = terawatt hours; 1 TWh = 1 billion kWh); EEG = Renewable Energy Act, as of April 1, 2000; SEG = Power Feed-in Law from 1.1.1991 to 31.03.2000; BGB = Construction Code (source: [4]).

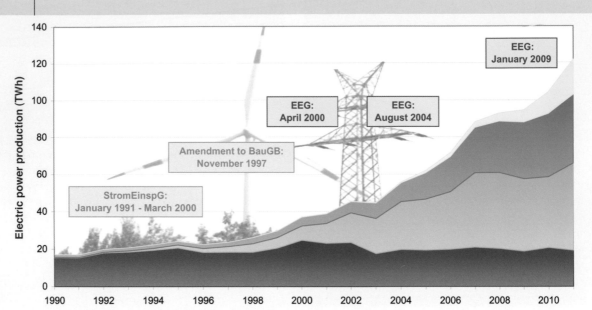

can lead to the growth of a whole new high-tech industry. Today, this industry is an economically successful global player. The Technical University in Berlin analyzed these developments over the past decades in a research project funded by the German Federal Ministry for the Environment [8,9]. Let us look at the developments in Germany more closely:

Renewable energy use has increased apace in Germany in the past years. In the year 2011, 20 % of the power from German grids came from renewable energy sources, nearly seven times as much as even in 1990 [4]. This was initially due to the successful development of wind energy, but in the meantime, there are important contributions from bioenergy and photovoltaics. With an overall energy input of 46.5 terawatt-hours (TWh) in 2011, wind power has clearly outdistanced the traditionally available hydroelectric power (which contributed 19.5 TWh in 2011). Electric power

production from bioenergy sources (including the biogenic portion of burned waste) moved up to second place in 2011 at around 37 TWh. Photovoltaic power generation also caught up rapidly, and in 2011, it already contributed 3 % of the overall power production, at 19 TWh. It has thus increased by a factor of 300 since the year 2000. Geothermal power production still plays only a minor role. Figure 1 shows the rapid development dynamics of electric power production from renewable energy sources in Germany. In the first half of 2011, the fraction of the total electric power supplied by renewable sources had already increased to around 20 % [1].

Germany has thus exceeded the goal for the proportion of energy supplied from renewable sources set by the Federal government only a few years ago – at that time, 12.5 % was the aim for the year 2010. This represents a great success for all those involved. The new resolution of the gov-

ABB. 2 | FINAL ENERGY USE IN GERMANY, 2011

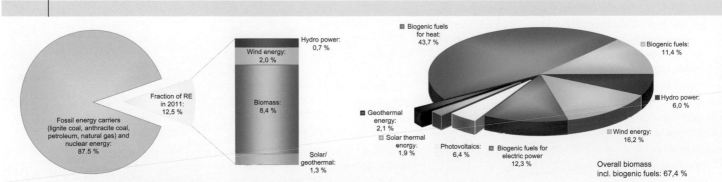

Left: The fractions of conventional and renewable energy sources(RE) within the overall final energy consumption in Germany; all together, 8,692 PJ (petajoule; 1 PJ = 10^15 J) was used in the year 2011. Right: The contributions from different renewable energy sources in 2011; all together they produced 300 TWh in 2011 (source: [4]).

ernment for the 'Energy Turnaround' (*Energiewende*), enacted on June 6, 2011, sets even more ambitious goals for the future use of renewable energy sources in Germany. For electric power, these new goals were already anchored in the amended EEG as of summer 2011 [1]. Its details are set out in the section 'Goals for Renewable Energy in Germany' on p. 11.

The Current Situation

Figure 2 (left) shows the distribution of the primary energy usage in Germany in the year 2010. It should not be surprising that fossil fuels still dominate the energy supply, providing together 89.1 % of the total [4]. Renewable energy sources had already attained a fraction of 10.9 % of the overall primary energy consumption by 2010. The right-hand part of Fig. 2 shows the origin of primary energy from renewable sources in the year 2010. Over two-thirds (71 %) of these renewable energy carriers are derived from the biomass. Wind energy contributes 13.4 %, water power 7.2 %, solar energy 6.3 % and geothermal energy 2.1 % (Figure 2).

The reason for the strong growth of renewable energy supplies in Germany is to be found mainly in political decisions. In the past twenty years, a public legal and economic framework was set up which has given renewable energy sources the chance to establish themselves on the market, in spite of their relatively high delivered power costs. Aside from various support programs and the market introduction program of the Federal government, the relevant laws were in particular the Power Feed-In Law (SEG) in 1990 and the Renewable Energy Act (EEG) in 2000, which gave the development of renewable energy sources an initial boost. The principle is straightforward: Power generated from renewable sources is given priority and a minimum price is guaranteed for power fed into the grid from these sources. On the basis of regular reports on the effectiveness of the EEG, the law is adjusted to the current situation as needed; most recently this was done in the summer of 2011 [1].

The prices paid for renewable-source power are scaled according to the source and other particular requirements of the individual energy carriers. They are graded regressively, i.e. they decrease from year to year. This is intended to force the renewable energy technologies to reduce their costs and to become competitive on the energy market in the medium term. The renewable energy technologies can accomplish this only through temporary subsidies, such as were given in the past to other energy technologies, e.g. nuclear energy. The renewable energy technologies will become strong pillars of the energy supply in the course of the 21st century only if they can demonstrate that they operate reliably in practice and are economically viable. To this end, each technology must go down the long road of research and development, past the pilot and demonstration plant stages, and finally become competitive on the energy market. This process requires public subsidies as well as a step-by-step inclusion of economic performance.

Potential and Limits

Often, the potential of the various technologies which exploit renewable energy sources is regarded with skepticism. Can renewable energies really make a decisive contribution towards satiating the increasing worldwide appetite for energy? Are there not physical, technical, ecological and infra-structural barriers to their increasing use?

Fundamentally, their potential is enormous. Most of the renewable energy resources are fed directly or indirectly from solar sources, and the sun supplies a continuous energy flux of over 1.3 kW/m^2 at the surface of the earth. Geothermal energy makes use of the heat from within the earth, which is fed by kinetic energy from the early stages of the earth's history and by radioactive decay processes (see the chapter "Energy from the Depths" in this book).

These energy sources are, however, far from being readily usable. Conversion processes, limited efficiencies and the required size of installations give rise to technical restrictions. In addition, there are limits due to the infra-structure, for example the local character of geothermal sources, limited transport radius for biogenic fuels, the availability of land and competition for its use. Not least, the limited availability and reliability of the energy supplies from fluctuating sources play a significant role. Furthermore, renewable energy sources should be ecologically compatible. Their requirements for land, potential damage to water sources and the protection of species, the landscape and the oceans set additional limits. All this means that the natural, global supply of potential renewable energy resources and the technically feasible energy production from each source differ widely (Figure 3).

In spite of these limitations, a widespread supply of renewable energies is possible. In order for it to be reliable and stable, it must be composed of the broadest possible mix of different renewable energy sources. In principle, water and wind power, use of the biomass, solar energy and

ABB. 3 | NATURAL SUPPLY AND AVAILABILITY

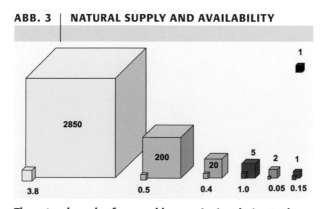

The natural supply of renewable energies in relation to the current world energy consumption (black cube, normalized to 1). Small cubes: The fraction of each energy source that is technically, economically and ecologically exploitable.
Yellow-green: solar radiation onto the continents; blue: wind; green: biomass; red: geothermal heat; violet: ocean/wave energy; dark blue: water power (source: [11]).

geothermal heat can together meet all of the demands. Germany is a good example of this. Although it is not located in the sunny South, and has only limited resources in the areas of hydroelectric and geothermal power, nevertheless renewable energy sources can in the long term supply all of Germany's energy requirements. Estimates put this contribution at around 800 TWh for electrical energy, 900 TWh for heat, and 90 TWh for fuels [4,11]. This corresponds to about 130 % of the current electric power consumption and 70 % of the current requirement for heating energy. With improved energy efficiency and a reasonable usage of power for heating and cooling as well as for transportation, the energy requirements in Germany can be met completely on the basis of renewable energy sources over the long term.

Water Power

Water is historically one of the oldest energy sources. Today, hydroelectric power in Germany comprises only a small contribution, which has remained stable for decades: 3 to 4 % of the electric power comes from storage and flowing water power plants. Its potential is rather limited in Germany, in contrast to the countries in the Alps such as Austria and Switzerland. In the future, it will therefore be possible to develop it further only to a limited extent. In 2010, the roughly 7000 large and small plants delivered about 20 TWh of energy, 90 % of this in Bavaria and Baden-Württemberg. The worldwide potential for hydroelectric power is considerably greater: nearly 16 % of the power generated

in 2010 came from hydroelectric plants [12,13]. Thus, water power – considered globally – is ahead of nuclear power. So far, it is the only renewable energy source which contributes on a large scale to the world's requirements for electrical energy. The other types of renewable energy sources contributed around 3 % to global electricity generation in 2010 [12,13]. In particular, 'large-scale water power' is significant. An example is the Chinese Three Gorge Project, which generates more than 18 GW of electric power, corresponding to about 14 nuclear power plant blocks (see the chapter "Flowing Energy").

In Germany, the so-called 'small-scale' water power still has limited possibilities for further development. New construction and modernization of this type of water power plants with output power under 1 MW however has ecological limits, since it makes use of small rivers and streams and it can affect their ecosystems. Synergetic effects can be expected when existing hydroelectric installations are modernized with transverse construction (dams) to increase their power generation capacities and at the same time to improve their hydro-ecological impacts. This development potential in Germany is estimated to imply an increase from currently 20 TWh up to 25 TWh per year.

The advantages of water power are obvious: The energy is normally available all the time, and water power plants have very long operating lifetimes. Furthermore, water turbines are extremely efficient, and can convert up to 90 % of the kinetic energy of the flowing water into electric power. By comparison: Modern natural gas combi-power plants

ABB. 4 | WIND POWER INSTALLATIONS IN GERMANY

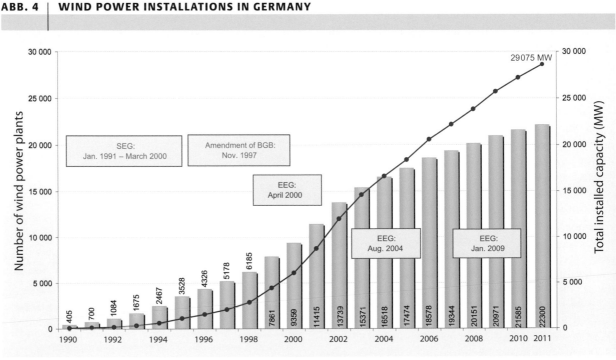

The development of wind energy e.g. in Germany from 1990 to 2011. The bars show the total number of wind power plants installed each year (accumulated); the blue curve gives the total installed generating capacity (right axis) (source: [14]).

have efficiencies of 60 %, and light-water reactors have only about 33 % efficiency.

Land-based Wind Energy

In Germany, the use of wind power (48.9 TWh) had clearly outstripped that of water power (18.8 TWh) by the year 2011. Modern wind energy plants attain efficiencies of up to 50 %. In 2011, plants yielding a wind power of about 2,000 MW were newly installed, bringing the total to 22,930 wind plants with an overall output power of 29,000 MW, generating about 7.6 % of the overall power consumed [14]. In the meantime, the so-called repowering is gaining momentum: Old plants are being replaced by more modern and more efficient installations. Thus, in 2011, 170 old plants with a nominal output power of 123 MW were replaced by 95 new ones with an overall output power of 238 MW [14].

In 2011, about 900 new plants were installed in Germany, with a total generating capacity of 2,000 MW; thus about 2.24 MW per installation. Given a long-term renewable potential wind power of 80,000 MW on land in Germany, and an average installed output power of 2.5 MW per plant, it would require 32,000 plants to realize the full potential of wind energy. At present, about 22,300 plants, each with an average power output of 1.3 MW, are in operation. Within the limitations of acceptance, citizen participation, questions of noise pollution, and the interests of nature and landscape conservation, it will be important in the coming years, in the course of authorization proceedings and land planning, to set up more efficient wind plants on higher towers (greater power yields!) at suitable locations.

This will permit the total number of plants to be limited, while at the same time increasing the overall yield: 32,000 plants on land, each with 2.5 MW output power, operating 2,500 full-power hours per year, would deliver a total of 200 TWh of electrical energy; that is about one-third of the current demand. This would be possible by making use of suitable sites on the seacoasts, but also in the interior by employing tower heights of over 100 m. A smaller portion of this potential could also be realized by installing smaller modern plants, taking the above criteria into account. A roughly equal potential of 200 TWh per year could in addition be realized by offshore wind plants in the Baltic and North Seas, so that simply by exploiting the available wind energy in Germany, in the long term, two-thirds of the current electric energy demand (of about 600 TWh/year) could be provided.

On windy days, the yield of wind energy in certain regions of Germany already exceeds the demand; on the other hand, on quiet days, other power sources have to compensate for the variable output of wind power plants. This applies increasingly also to photovoltaic plants, while in contrast, hydroelectric plants and the biomass have a 'built-in' storage capacity and can thus be independently regulated to meet demand. The future energy supply system will

ABB. 5 | FUTURE POWER GENERATION

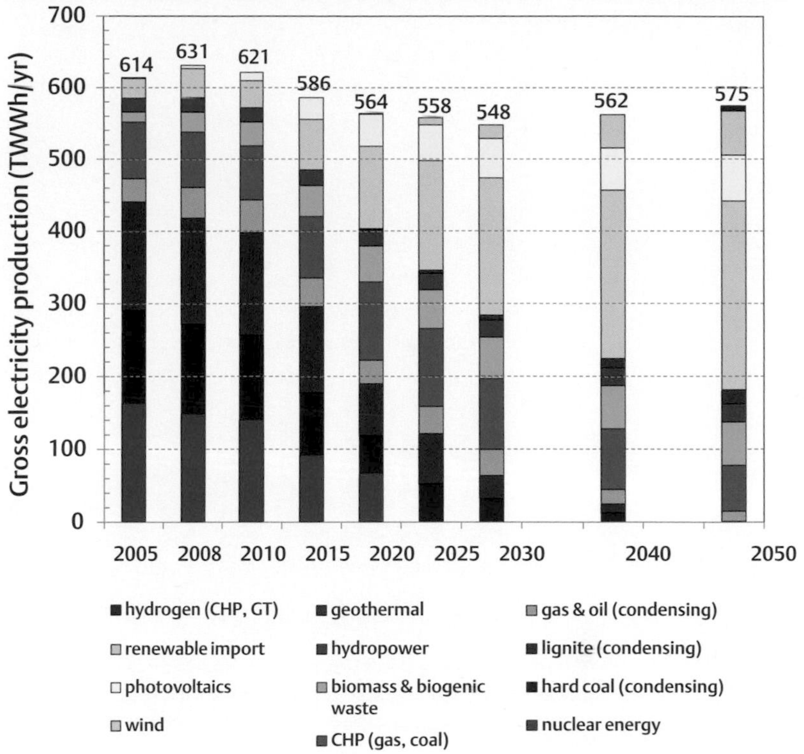

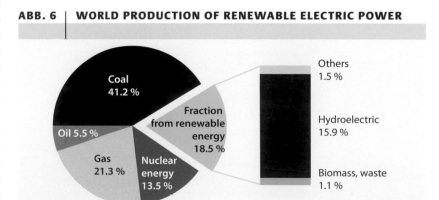

Electric power production in Germany according to type of power plant and energy source, from the long-term scenario "Lead Study 2011". The numbers above the bars give the total energy generated in TWh. The nuclear energy exit scenario corresponds to the Exit Resolution of June 6, 2011. RE = renewable energy (source: [16]).

have to be able to deal with the fluctuating supply of power by means of rapidly controllable, decentral power plants (CHP, natural gas or gas produced using renewable energy), energy storage reservoirs, power management, etc. This new orientation of power system optimization based on supply security and making use of control engineering, information

ABB. 6 | WORLD PRODUCTION OF RENEWABLE ELECTRIC POWER

The fractions of electric power produced from various energy sources in 2008 (sources: [4,18]).

ABB. 7 | WORLD POPULATION AND PRIMARY ENERGY CONSUMPTION

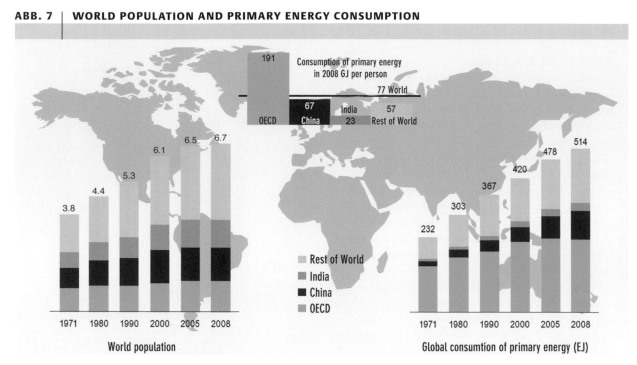

The growth of the world population (in billions) and its consumption of primary energy (EJ = exajoule, 10^{18} J). In 2008, each person in the OECD countries consumed on the average 191 GJ (GJ = 10^9 J), in China 67 GJ, in India 23 GJ, and in the rest of the world, 57 GJ. The average energy consumption worldwide was 77 GJ per person (source: [18]).

and communications technology can be considered to be a major challenge for the coming years [15,16].

Biomass

The utilization of energy from the biomass is often underestimated. At present, biogenic heating fuels are being rediscovered in Germany. Wood, biowastes, liquid manure and other materials originating from plants and animals can be used for heating and also for electric power generation. The combination of the two uses is particularly efficient. In Germany, currently 90 % of the renewable heat energy originates from biofuels, mainly from wood burning – but increasingly also from wood waste, wood-chip and pellet heating and biogas plants, as well as the biogenic component of waste. Its contribution to electric power generation is also increasing: in 2011, it was 6 % of the overall German demand, corresponding to 37 TWh.

Biofuels are available around the clock and can be utilized in power plants like any other fuel. Biogenic vehicle fuels, as mentioned above, are getting renewable energy carriers rolling as suppliers for transportation.

Biogenic fuels, however, have come under massive public criticism, because they are not always produced under ecologically and socially acceptable conditions. In the worst case, they can even yield a poorer climate balance than fossil fuels. They thus require a detailed critical analysis and optimization process for each product, as is discussed in detail in the chapter "Biofuels: Green Opportunity or Danger?".

Solar Energy

Solar energy is *the* renewable energy source par excellence. Its simplest form is the use of solar heat from collectors, increasingly employed for household warm water heating and for public spaces such as sports halls and swimming pools. More than 15 million square meters of collectors were installed on German rooftops as of 2011 [14].

Solar thermal power generation has meanwhile also made the transition to commercial applications on a large scale (see also the chapter "How the Sun gets into the Power Plant"). Parabolic trough collectors, solar towers or paraboloid dish reflector installations can produce temperatures of over 1000 °C, which with the aid of gas or steam turbines can be converted into electric power. These technologies could in the medium term contribute appreciably to the electric power supply. They are however efficient only in locations with a high level of insolation, such as in the whole Mediterranean region. Germany would thus have to import solar power from solar thermal plants via the common power grid, which initially could be laid out on a European basis; in the long term, North African countries could supply solar power via a ring transmission line around the Mediterranean Sea (see also the chapter "Power from the Desert") [11,16,17].

The most immediate and technologically attractive use of solar energy is certainly photovoltaic conversion. The market for photovoltaic installations currently shows the most dynamic growth: Between 2000 (76 MW) and 2011 (25,039 MW), the installed peak power capacity increased in Germany by a factor of more than 300. This corresponds to a growth rate of 72 % per year during the past decade [4]. New production techniques at the same time offer the chance to produce solar cells considerably more cheaply and with less energy investment, and thus to allow a breakthrough onto the market (see the chapters "Solar Cells – An Overview", "Solar Cells from Ribbon Silicon", "Low-priced Modules for Solar Construction", and "On the Path towards Power-Grid Parity").

Geothermal Energy

The renewable energy resource which at present is the least developed is geothermal heat. Deep-well geothermal energy makes use either of hot water from the depths of the earth, or it utilizes hydraulic stimulation to inject water into hot, dry rock strata (hot-dry rock process), with wells of up to 5 km deep (see the chapter "Energy from the Depths"). At temperatures over 100 °C, electric power can also be produced – in Germany for example at the Neustadt-Glewe site in Mecklenburg-Vorpommern. Favorable regions with high thermal gradients are in particular the North German Plain, the North Alpine Molasse Basin, and the Upper Rhine Graben.

Geothermal heat has the advantage that it is available around the clock. Nevertheless, the use of geothermal heat and power production is still in its infancy. Especially the exploitation of deep-well geothermal energy is technically challenging and still requires intense research and development. Near-surface geothermal energy is more highly developed; heat pumps have long been in use.

The exploitation of deep-well and near-surface geothermal heat more than tripled in the decade from 2000 (1.7 TWh) to 2010 (5.6 TWh). If and when it becomes possible to utilize geothermal energy on a major scale, then its constancy and reliability will make a considerable contribution to the overall energy supply. Its long-term potential in Germany is estimated to be 90 TWh/year for electric power generation and 300 TWh/year for heating.

The Window of Opportunity

How will energy supplies in Germany develop in the future? Will all the renewable energy source options play a role, and if so, to what extent? The resolution of the federal government on June 6, 2011 contains the following elements for an energy turnaround in Germany:

- An exit strategy for nuclear power in Germany by the end of 2022;
- Continuous development of the use of renewable energy sources;
- Modernization and further development of the electric power grid;
- Energy conservation and an increase in efficiency in all areas concerning energy;
- Attaining the challenging climate protection goals and thereby a clear-cut reduction in the consumption of fossil fuels.

DEVELOPMENT GOALS FOR RENEWABLE ENERGY IN GERMANY

The German Federal government, with its energy Lurnaround (*Energiewende*) package adopted on June 6th, 2011 and the amendment of the Renewable Energy Act, is pursuing the goals set out here: The fraction of renewable energy sources in electric power generation are to increase as follows:
- by 2020 at the latest up to at least 35 %
- by 2030 at the latest: up to at least 50 %
- by 2040 at the latest: up to at least 65 %
- by 2050 at the latest: up to at least 80 %

The goals for growth of the fraction of renewable energy sources in overall energy consumption (electric power, heating/cooling, transportation) are:
- by 2020: 18 % (corresponds to the EU directive; see above)
- by 2030: 30 %
- by 2040: 45 %
- by 2050: 60 %

Furthermore, by 2020 their contribution to space heating in total should increase to 14 % and their contribution to energy use in the transportation sector to 10 %.

The Federal cabinet also enacted additional goals in Berlin on June 6th, 2011, to which the development of renewable energy sources makes essential contributions. The German emissions of greenhouse gases are to be decreased by 40 % by 2020, based on the reference year 1990, and by 80 to 95 % by 2050. Consumption of electric power is to decrease by 10 % up to 2020 and by 25 % up to 2050; consumption of all primary energies by 20 % up to 2020 and by 50 % up to 2050.

The goal is a transition to a secure energy supply based for the most part on renewable energy sources in the long term. The basis for such a transition was already laid down in the past decade, with a renewable electric power fraction of 20 %, and 12 % of the overall energy supply in 2011. The upcoming system transformation will require continuing strong commitment and efforts.

The fact that the development of renewable energy sources is already leading to a number of positive results, including economic effects, is shown by its achievements up to the year 2011:

- Reduction of greenhouse gas emissions by 130 million tons;
- Avoided environmental damage worth about 10 billion € (especially climate damage at an average value of 80 €/ton of CO_2);
- Reduced imports of energy carriers: ca. 6.5 billion €;
- Investments: 22.9 billion €;
- Employment: 381,600 jobs;
- Increase of regional added value.

If all the relevant quantities are considered (systems-analytical cost-benefit effects, distribution effects, macroeconomic effects), the benefits today already outweigh the costs. Nevertheless, support will still be necessary in the foreseeable future, since these quantities are related in a complex way [4,19,20]. In the course of the cost regression for subsidies of the various technologies making use of renewable energy sources, and the expected cost increases for fossil energy carriers due to their limited supplies and harmful effects on the climate, the beneficial aspects of renewable energy sources will presumably become more and more apparent [16].

Offshore and the Open Field

The next major step in the modification of the energy systems in Germany will be the start-up of offshore wind energy. Along the German seacoasts and within the 'exclusive economic zone' (EEZ), which extends out to a distance of 200 nautical miles (370km) from the coastline, a potential power-generating capacity of up to 25 GW of electric power output is predicted by the year 2020.

Such offshore wind installations will have to be built far from the coastline in water depths of up to 60 m. This is particularly true of the North Sea, which has strong winds. In the shallow water near the coasts, there are no suitable sites due to nature conservation areas, traditional exploitation rights such as gravel production, restricted military zones and ship traffic. Plants in deeper water, however, require a more complex technology and are more expensive. The high-power sea cables for transmitting the power to the coast over distances of 30 to 80 km will also drive up the investment costs.

However, the offshore installations far from the coast have a considerable advantage: The wind from the free water surface is stronger and steadier. This compensates to some extent for the higher costs of these wind parks. To be sure, the individual plants must deliver high power outputs. Only when they achieve an output power capacity of at least 5 MW_{el} can they be economically operated under such conditions. A pioneering role in this development is being played by the wind park *Alpha Ventus*, which stands in water 30 m deep and 45 km in front of the coast of the island of Borkum: On August 12, 2009, the first 5 MW wind energy plants started delivering power, and in the meantime, all 12 plants are in operation [21]. The Fino offshore platforms perform useful services for the development of offshore wind parks. The Fino Research Initiative in the North and Baltic Seas is financed by an Offshore Trust, founded by commercial firms, nonprofit organizations and power-grid operators, and supported by the Federal Ministry for the Environment [22].

Scenarios for Ecologically Optimized Development

Just how the proportion of renewable energy sources within the energy mix in Germany will evolve in reality cannot of course be precisely predicted. However, model calculations make it clear which paths this evolution might take under plausible assumptions. The Institute for Technical Thermodynamics at the DLR in Stuttgart carried out a comprehensive study in 2004, analyzing various scenarios [23]. They took into account technical developments, economic feasibility, supply security and ecological and social compatibility. This study illustrates the essential trends. A series of other studies on ecological optimization and accompa-

nying research has looked into various individual renewable energy technologies.

Figure 5 gives the distribution of power generation in Germany according to the type of power plant and the energy source within the long-term scenarios 2011 of the "Lead Study 2011" [16]. These scenarios aim at an economically acceptable increase in the use of renewable energy sources, but also take ecological factors into account. For over twelve years, the Federal Ministry for the Environment has issued such scenarios for the development of renewable energy sources. These scenarios have considered development paths which are ecologically optimized and are designed around sustainability criteria. They consider the dynamics of technical and economic developments and the interactions of the whole energy system in view of increasing contributions from renewable energy sources.

The so-called "Lead Study 2011" [16] took into account the energy turnaround package of the Federal government, in which nuclear energy is to be phased out by 2022. All of its assumptions agree precisely with the Resolution of June 6th, 2011, and they still represent a very felicitous summary of the development of renewable energy technologies, of other energy carriers, and of the necessary transformation of the overall energy system. This study also shows clearly that the required reductions in greenhouse gas emissions by 2020 and 2050 can in fact be accomplished: Half of the reductions through the continued development of renewable energy sources, and the other half through reduced energy consumption, improved energy efficiency and the reduction of the consumption of fossil energy carriers, in spite of the phasing-out of nuclear power.

Renewable Energy on a Worldwide Scale

Figure 6 shows the contributions of various energy carriers to worldwide electric power generation in 2008. Fossil fuels were predominant, at 68 % of the total, while renewable energy sources already supplied 18.5 %, and nuclear energy 13.5 %. In the areas of heating and transportation, biogenic fuels in particular supply an appreciable fraction, which however must be critically examined in terms of its real sustainability.

Renewable energy sources can also play the leading role in the long-term global energy supply [12,13,24]. However, their further development alone will not achieve this goal. Thus, Figure 7 shows the parallel increase of the world's population and of the global energy demand from 1971 to 2008 [18]. Without an energy turnaround on a global scale, reversing these trends will not be possible. We can reach the goal of a global energy supply with a high proportion of energy from renewable sources on a long-term basis only if we make additional strong efforts. One of these concerns improved energy efficiency and access to energy. In addition, worldwide population growth must be slowed considerably.

Summary

By the year 2011, already 12,5 % of the final energy con-
sumption in Germany was supplied from renewable energy
sources; for electrical energy, the proportion was 20,3 %,
while for heating, it was 11 %, and for vehicle fuels, around
5.5 %. In the first half of 2012, its contribution to electric pow-
er generation had already risen to ca. 25 %. The German Fed-
eral government, with its resolutions of June 6th, 2011, in-
tends (in the energy turnaround – Energiewende) to secure a
continuous further development, which satisfies all of the eco-
logical, economic and social criteria of sustainability. In Ger-
many, a productive industrial sector with nearly 400,000 em-
ployees has developed around the exploitation of renewable
energy sources. The goals enacted by the government are am-
bitious: at least 35 % of electric power to come from renew-
able sources by 2020 at the latest, and at least 80 % by 2050
at the latest; 18 % of the overall energy consumption by 2020,
and 60 % by 2050. This national strategy is embedded in an
EU Directive for the advancement of renewable energy
sources. For global energy supplies, also, renewable sources
must assume the predominant role. Successes within the
EU and in other countries can serve as examples. Worldwide,
18.5 % of the electric power was generated from renewable
sources.

References

[1] German Federal Ministry for the Environment, BMU: Web pages on renewable energy, www.erneuerbare-energien.de/english/ renewable_energy/aktuell/3860.php.

[2] EP/ER: Directive of the European Parliament and Council, 2009/28/EG from April 23rd, **2009**, for *The Advancement of the Use of Energy from Renewable Sources*, Official Register of the EU, L140/15 June 2009.

[3] International Feed-In Cooperation, www.feed-in-cooperation.org.

[4] BMU – *Renewable Energy in Figures*, Brochure, August **2012**; available as pdf from www.erneuerbare-energien.de/english/ renewable_energy_in_figures/doc/5996.php.

[5] European Commission: Communication 31.1.**2011**: *Renewable Energy: Progressing towards the 2020 target*. Available from: bit.ly/TRPt5V.

[6] Eurostat, Statistical Office of the EU, Luxemburg: Online Database. See epp.eurostat.ec.europa.eu/portal/page/portal/energy.

[7] EWEA – Annual Report 2010 of the European Wind Energy Associa-tion, 2011. Download: www.ewea.org/index.php?id=11.

[8] E. Bruns *et al.*, *Renewable Energies in Germany's Electricity Market*; Springer, Heidelberg **2010**.

[9] Agency for Renewable Energies (Eds.): *20 Years of Support for Power from Renewable Energy in Germany*, See: www.unendlich-viel-energie.de/en/homepage.html.

[10] Bundesverband Windenergie (BWE), www.windenergie.de (in Ger-man); European Wind Energy Association (EWEA), www.ewea.org; Global Wind Energy Council (GWEC), www.gwec.net.

[11] BMU – *Renewable Energies – Perspectives for a Renewable Energy Future*, Brochure, Heidelberg, **2011**; See: www.erneuerbare-energien .de/english/renewable_energy/downloads/doc/44744.php.

[12] International Renewable Energy Agency (IRENA), **2011**. www.irena.org.

[13] Renewable Energy Policy Network – REN21: *Renewables 2011 Status Report*: www.ren21.net.

[14] German Wind Energy Institute (DEWI): DEWI **2011**: *Jahresbilanz Windenergie 2010*. See: www.dewi.de/dewi/index.php?id=1&L=0.

[15] Fraunhofer Institute for Wind Energy and Energy Systems Technolo-gy, *IWES 2011*; See: www.iwes.fraunhofer.de/en.html.

[16] DLR, IWES, IfnE: Long-term scenarios 2011, "*Lead Study 2011*", commissioned by the BMU, March **2012**. See: www.erneuerbare-energien.de/english/renewable_energy/downloads/doc/48532.php.

[17] Desertec Foundation 2011, See: www.desertec.org.

[18] International Energy Agency (IEA), *Renewables Information*, Edition 2010. IEA/OECD, Paris **2010**.

[19] ISI, GWS, IZES, DIW: *Individual and Global Economic Analysis of Costs and Side Effects of the Development of Renewable Energies on the German Electric Power Market*, Update **2010**. See: www.erneuerbare-energien.de/english/renewable_energy/studies/doc/42455.php.

[20] GWS, DIW, DLR, ISI, ZSW: *Short- and Long-Term Effects on the Employment Market of the Development of Renewable Energies in Germany*, commissioned by the BMU (Ed.), Feb. **2011**, See: www.erneuerbare-energien.de/english/renewable_energy/ studies/doc/42455.php.

[21] *Alpha Ventus 2011*, www.alpha-ventus.de/index.php?id=80.

[22] Fino Offshore Platforms **2011**. See: www.fino-offshore.de (in German).

[23] Nitsch *et al.*: *Ecologically optimised development of the utilisation of renewable energies*, DLR: Stuttgart **2004** (in German). Commis-sioned by the BMU. See: www.erneuerbare-energien.de/english/ renewable_energy/studies/doc/42455.php.

[24] IPCC *Special Report on Renewable Energies 2011*. See: www.ipcc.ch.

The publications of the BMU can be ordered from the Department of Public Relations (Oeffentlichkeitsarbeit) in Berlin or from www.erneuer-bare-energien.de.

About the Authors

Harald Kohl studied physics in Heidelberg and carried out his doctoral work at the Max-Planck Institute for Nuclear Physics there. Since 1992, he has worked at the Federal Ministry for the Environ-ment, Natural Conservation and Nuclear Safety (BMU) in Bonn and Berlin. He is currently head of the Division of Public Information.

Wolfhart Dürrschmidt studied physics in Tübingen and earned his doctorate at the Institute for Physical and Theoretical Chemistry there. He is head of the Division of Fundamentals and Strategy for Renew-able Energy at the BMU in Berlin.

Addresses:
Dr. Harald Kohl, Bundesministerium für Umwelt, Naturschutz und Reaktorsicherheit (BMU), Referat K, Stresemannstr. 128–130, 10117 Berlin, Germany.
Dr. Wolfhart Dürrschmidt, BMU, Referatsleiter KI III 1, Renewable Energies
Köthener Str. 2–3, 10963 Berlin, Germany.
harald.kohl@bmu.bund.de
wolfhart.duerrschmidt@bmu.bund.de

Wind Energy

A Tailwind for Sustainable Technology

by Martin Kühn | Tobias Klaus

In Germany, more than 22,000 wind-energy plants are now online, providing about 10 % of the total power consumption. They have thus outstripped every other sustainable energy form here [1]. The Federal Ministry for the Environment considers a contribution of 25 % by the year 2030 to be possible. What potential does wind energy still hold?

Mankind has been making use of wind power for at least 4000 years. In Mesopotamia, Afghanistan and China, wind-powered water pumps and grinding mills were developed very early, apart from to the use of wind power for sailing ships. In earliest times, windmills utilized a vertical-shaft rotor, which was driven by the drag force acting on the rotor blades by the wind. This design concept, known as a drag device, has a low efficiency, roughly a fourth of that of the aerodynamic rotors described in the following sections [2]. It is still used by the widespread cup anemometers that measure wind velocity.

In Europe from around the 12th century onwards, new windmill types were developed, such as the post windmill, the tower mill, and later the Dutch windmill. They were introduced to provide an important complement to human or animal muscle power. The decisive advance in these historical windmills in the western world was not the generally horizontal orientation of their rotor shafts, but rather the fact that the flowing air has a higher velocity at the rotor blades and drives them via the aerodynamic lift force,

perpendicular to the flow velocity. For a drag device, which is moving with the flow, the relative velocity at the rotor blades is always lower than the wind velocity itself. Lift devices, in contrast, can achieve higher apparent wind speeds by vector addition of the wind velocity and the circumferential velocity of the rotor. Only in this way can the forces necessary for an optimal deceleration of the wind be generated, and the aerodynamic efficiency approaches its theoretical maximum of 59 % [2].

The best-known examples of these machines were the four-bladed Dutch windmill and the 'Western mill', which turned slowly and was used to pump water, with twenty or more rotor blades. The latter, developed in mid-19th century America, was the first wind-powered device to be produced industrially on a large scale. It was able to operate in automatic mode without human attention. A robust control system with two weather vanes kept the rotor pointed towards the wind, and turned it away if the wind became too strong, to avoid damage from overload.

Three-bladed Turbines with High Tip-Speed Ratio

The invention of the steam engine and later of electric motors during the Industrial Revolutions led to a decline in the use of windmills as working machines. Only the Western windmills were still used to some extent as decentralized water pumps. The Dane Paul La Cour was the first, in 1891, to develop a windmill for generating electricity. He recognized the fact that along with increasing the aerodynamic efficiency, it was also favorable for the construction if the circumferential velocity of the blades were considerably higher than the wind velocity. In these turbines with a high tip-speed ratio, only a few very slim blades are required, and the generator is driven at a relatively high rotational speed with a correspondingly low torque. Albert Betz, Frederick W. Lancaster, and Nikolai J. Joukowski generalized these findings in parallel to each other and derived the maximum attainable aerodynamic efficiency of 59 %.

Every wind-power installation requires a method of controlling the energy input and the load on the plant, since the energy transferred from the wind increases as the third power of its velocity. Two principles have established them-

Renewable Energy. Edited by R. Wengenmayr, Th. Bührke. Copyright © 2013 WILEY-VCH Verlag GmbH & Co. KGaA, Weinheim

The Danish offshore windpark Horns Rev consists of 80 units, each with 2 MW output power, located 14 to 19 km north-west of Esbjerg in the North Sea. The ocean is 5 to 15 m deep here (photo: Vestas Central Europe).

selves for providing this control mechanism: stall and pitch. They were developed beginning with La Cour and continuing through the work of wind-energy pioneers in Denmark, France, the USA and Germany.

In the simplest design (stall), the rotor blades are rigidly attached to the hub (Figure 1). The rotational speed is held nearly constant by an asynchronous generator coupled to the power grid. This is typically a three-phase motor operated in generator mode. When the wind becomes stronger, its angle of attack on the rotor blades changes as a result of the vector addition of the wind velocity and the circumferential velocity of the rotor. This increase in angle of attack leads to a flow separation on the low-pressure side of the blades, and thus to a stall. This protects the wind turbine from excessive power intake, since the lift acting on the blades is reduced and their drag is increased (Figure 2).

This simple and robust system was introduced in 1957 by the Danish wind-power pioneer Johannes Juul. Due to its country of origin, it is known as the 'Danish Concept'. It was important for the early deployment of wind-energy installations in large numbers in the mid-1980's, with rotor diameters of 15 to 20 m and output power of 50 to 100 kW. In the following decade, the principle was developed further into the 'active-stall concept'. In this construction, the stall effect can be actively induced: By varying the pitch of the blades, i.e. increasing the angle of attack by a few degrees (turning the trailing edge into the wind), the flow

separation can be actively controlled and the desired effective power can be reliably regulated.

The second principle for limiting power intake is based on a greater variation of the rotor blade angle, or pitch. If the wind speed increases after the nominal power capacity has been reached, then the leading edge of the rotor

FIG. 1 | A WIND-POWER INSTALLATION

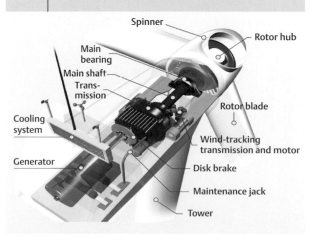

The construction of a stall-regulated wind-power installation with a transmission and constant rotation speed, designed by **NEG-Micon** *(graphics: Bundesverband Windenergie).*

FIG. 2 | THE STALL CONCEPT

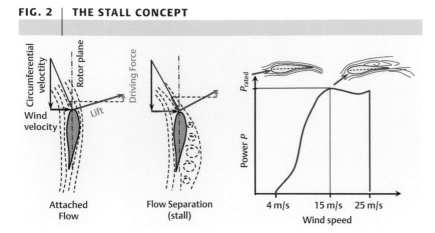

Left: Control of power uptake with increasing wind velocity by flow separation or stalling; right: The power-uptake curve, showing the limiting of power uptake by stalling.

blades is turned into the wind (Figure 3). By decreasing the angle of attack, the power and the load are reduced.

This concept, oriented towards lightweight construction, was decisively influenced by the German wind-energy pioneer Ulrich Hütter in Stuttgart. In 1957, he constructed a pitch-regulated two-blade device, in which for the first time rotor blades made of fiberglass-reinforced plastic were used [3]. This construction method became standard from the 1980's on. At the time, it was the first application of a new fabrication material for such large structural components. Only later were applications in aeronautics and other areas of industry introduced.

From Grid-connected to Grid-supporting Wind Power Plants

Even though the external appearance of wind-power installations has not changed much in the past 20 years, a rapid technical development has taken place, which is not outwardly apparent: Increasingly, larger and more efficient

FIG. 3 | THE PITCH CONCEPT

Left: Power control using pitch regulation; Right: The power curve.

wind turbines are feeding electrical energy of improved quality and at lower cost into the power grid. A decisive factor was the introduction of operation at a variable rotational velocity, leading to the devices now called wind turbines.

It soon became clear that plants which operate at a constant rotational velocity cannot completely compensate a gusty wind even if the rotor blades are adjusted very quickly, and that those plants were thus subject to strong short-term variations in output power, accompanied by both structural loads and corresponding reactions on the power grid. The advantages of the pitch concept, i.e. a constant nominal output power and good performance during startup and during storms, can be put into practice only in combination with a certain variability in the rotational velocity of the turbine. This however requires some additional effort in the design of the electrical components. To this end, from initially three, two types of construction have become common.

At first, especially the Danish firm *Vestas* introduced a process that allows the variation of the rotational velocity by up to ten percent. This is accomplished by a fast regulation of the rotational-velocity compliance (slip) of the asynchronous generator, which is coupled to the power grid. Through the interactions of the rotor, which now acts as a flywheel, with the somewhat slower pitch adjustment, wind variations above the nominal operating speed can be smoothed out very satisfactorily.

Mainly in Germany, beginning in the 1980's with experimental installations and commercially from 1995 on, a concept involving complete variability of the rotational velocity was developed, which today is used in more than half of all new plants. While the stator of the asynchronous generator is still coupled directly to the power grid, the rotor accepts or outputs precisely the AC frequency which is required to adapt to the desired rotational velocity. By means of such a doubly-fed asynchronous generator, the rotational velocity can be roughly doubled between the startup speed of about 3.5 m/s and the nominal operating speed of 11 to 13 m/s. The rotor functions near its aerodynamic optimum, and aerodynamic noise is effectively reduced. Above the nominal operating speed, the rotational velocity oscillates by ca. ±10 %, in order to smooth out wind gusts, again in combination with the pitch adjustment.

The most evident, but complex path to complete variability of the rotational velocity lies in electrically decoupling the generator using a transverter, via an intermediate DC circuit. In this concept, in which as a rule a synchronous generator is employed, all of the electrical power is passed through the frequency transverter. By controlling the excitation in the generator rotor, the rotational velocity can be varied by up to three times its startup value. The *Enercon* company, market leader in Germany, applies this concept very successfully to gearless wind-energy plants, using a specially-developed direct-drive multipole synchronous generator (Figure 4). In recent years, this principle, owing to its excellent grid compatibility and its independence of the

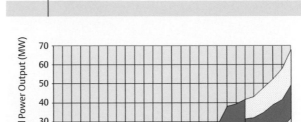

FIG. 4 | ADJUSTABLE-PITCH PLANTS

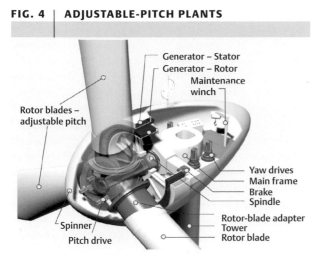

Generator – Stator
Generator – Rotor
Maintenance winch
Rotor blades – adjustable pitch
Yaw drives
Main frame
Brake
Spindle
Rotor-blade adapter
Tower
Rotor blade
Spinner
Pitch drive

Internal construction of a variable-speed, adjustable-pitch wind-energy plant without a transmission, built by the Enercon company (graphics: Bundesverband Windenergie).

FIG. 5 | HISTORY

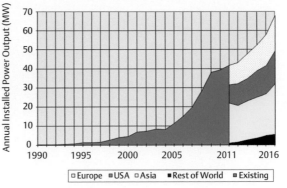

Annual wind-power evolution over time – current data, 1990-2011, and forecast for 2012–2016 (graphics: BTM Consult – A part of Navigant).

local grid frequency, has also been applied to some transmission-based machines, which still supply ca. 85 % of the world market.

In the meantime, the latter two concepts of adjustable pitch and variable rotational velocity have asserted themselves in the market and have practically superseded the simple, robust stall-regulated plants of earlier days. Partial or complete decoupling of the generator from the power grid provides a great improvement in grid compatibility and even – under favorable circumstances – allows the support of the electrical power grid. The phase angle between the current and the voltage (power factor) can be adjusted as needed. Negative effects on the grid, such as switching currents, voltage and power variations and harmonics, can be avoided or greatly reduced. Furthermore, the installations are much less sensitive towards disturbances from the grid, such as temporary voltage breakdowns.

Lightweight Construction, Intelligent Installations, and Reliability

Today's wind power plants, with rotor diameters of up to 127 m and a nominal output power of up to 7.5 MW, are among the largest rotating machines in existence. They defy the extremely harsh environmental conditions in the atmospheric surface layers near the ground by employing complex automatic control systems, for example by monitoring a number of different operating parameters or by using laser optical-fiber load sensors in the rotor blades. Furthermore, the most modern structural materials are used, such as carbon-fiber composites or dynamically-tough cast and forged alloys.

Due to the temporal and spatial structure of wind gusts, every local flurry has a multiple effect on the rotating blades. Within the planned lifetime of twenty years for a wind-energy plant, up to a billion load cycles occur – an order of magnitude completely unknown in other areas of

mechanical engineering. At the same time, the increasing size of the plants requires more lightweight construction methods; otherwise, the materials stresses resulting from the continual alternating bending forces from the rotor blades' own weight would become a problem.

In terms of commercial competitiveness compared to conventional power plants, cost savings also dictate the technical developments. They must be achieved not only by economies of scale through mass production of large numbers of plants, but also by increasing the efficiency of the individual plants. Frequently, the maximum theoretical aerodynamic efficiency is already approached quite closely; therefore, one tries to further reduce the investment costs per kilowatt hour generated. This can be achieved for example through active and passive vibration damping, compensation of variable loads, and the application of lightweight construction concepts. In addition, the operating costs can be decreased by a further improvement in the reliability of the installations.

The technical availability of installations, i.e. the fraction of the time during which the turbines are operable, is in the meantime near 98–99 % [2]. Nevertheless, further improvements in the durability of expensive components such as rotor blades and transmission, and in the reliability of electrical components and sensors, are necessary. This applies in particular to plants in the megawatt class. Such plants have been installed in large numbers since the end of the 1990's and at the beginning of the past decade, often after only an all-too-brief try-out period.

Wind Energy in the Updraft – Offshore Plants

In recent years, wind energy has experienced a worldwide boom. Up to the end of 2011, on a global scale, plants for nearly 239,000 MW were installed; 42,000 MW of this within 2011 alone. The world market, in which German manufacturers of plants and components have a share of more

Fig. 6 *Installation of an offshore wind energy plant with 5 MW output power, off the Scottish coast in August 2006. The diameter of the rotor is 126 m* (photo: REpower System AG).

than 17 % of the total added value, is growing annually at a rate of over 20 % (Figure 5) [4,5].

With an annual turnaround of nearly 5 billion €, the export sector of the German plant and component manufacturers represents about 66 % of their total production (2010). Even though Germany is in the meantime no longer the most important market, continued expansion is taking place in other European countries, in the USA, and in the emerging Asian markets, particularly in the PR China and India. Wind energy has evolved into a non-negligible component of the global energy system, in which German industry continues to play a leading role. With the increasing growth of these markets, questions of the exploitation of the enormous wind resources on the oceans, integration into the international energy system, economic returns, nature and landscape protection, and not least social acceptance, are growing more pressing.

In near-coastal regions of the oceans, there are enormous wind resources waiting to be tapped. Besides a higher energy yield by 40 to 50 % compared to good onshore sites, a greater site area is available here. The Federal Environment Ministry in Germany in 2010 predicted the installation of 25 GW from offshore plants by 2030, covering 15 % of the German power requirements. In a first step, the Federal government plans to increase onshore generating capacity from 27 to 36 GW by 2020, and offshore capacity from 0.2 GW to 10 GW in the same time period [6].

Following the first suggestions for offshore wind projects in the 1970's, during the 1990's several smaller European demonstration projects were set up. After 2000, the construction of commercial wind parks with up to 160 MW output power was begun, using individual plants in the 1.5

to 2 MW class. By late 2008, the installed offshore power output totalled nearly 1,500 MW. This corresponded to about 1.2 % of the worldwide installed electric generating capacity from wind energy. Operating experience has thus far been mainly positive and supports further development, which presently is taking place, in particular in Great Britain, Denmark, the Netherlands and Sweden.

As with any new technology, there were also setbacks. In mid-2004, at the largest Danish offshore wind park at Horns Rev, only two years after completion of its construction, all eighty plants had to be temporarily taken down and overhauled on land at considerable expense – the transformers and generators were not sufficiently protected against the harsh saltwater environment. This however also demonstrated that by now, the industry is sufficiently mature to survive impacts of this magnitude; by mid-December of the same year, all the plants were again on line.

In Germany, the water depths of 25 to 40 m and offshore distances of 30 to over 100 km in suitable areas represent a financial hurdle for the initial projects, in particular. The first 'genuine' offshore project in Germany is the test field *Alpha Ventus*, 45 km north of the island of Borkum, which was completed at the end of 2009. There, twelve wind-energy plants of the currently most powerful 5 MW class are in operation; only four German manufacturers offer plants of this size. In 2006, a plant of this type was installed on a cantilevered foundation in a water depth of 44 m off the Scottish coast (Figure 6). All over the world, the construction of additional offshore parks has been authorized.

For the future development of wind energy, differing predictions have been made. The European Wind Energy Agency (EWEA) expects an increase in the overall installed power from 3 GW (2010) to roughly 9 GW in 2013 and 40 GW by 2020. By 2030, according to this prognosis, offshore installations with an output power of 150 GW [7] will be on line. The most important markets are expected to be Great Britain and Germany. The Danish firm *BTM Consult* predicts for the year 2014 a worldwide total offshore wind power capacity of 16 GW, most of which is expected to be in Europe. The strongest growth in the foreseeable future will be on land, so that the fraction of offshore wind energy relative to the overall installed output power is estimated to be 10 % in the year 2015 [7].

Grid Integration in Spite of Varying Power Outputs

In general, it is expected that a proportion of up to 20 % of renewable energy sources such as wind power and solar power can be integrated into the power grid without major problems. Following the decision of the German Federal government to shut down successively all the nuclear power plants by 2022, the integration of new plants into the grid represents a technical and economic challenge. The fraction of power from sustainable sources is expected to increase from 20 % in 2011 to 35 % by 2020, in order to decrease the emissions of greenhouse gases relative to 1990 by 40 %.

The particular challenge for wind energy is due to the regional concentration of wind plants in the Northern and Eastern coastal areas, and to the daily and seasonal variations of the wind. At times the input of wind energy there exceeds the local grid demand, while at other times, there is almost no wind power available.

Decentralized power inputs, e.g. of wind power into the weak periphery of the power grid, new power-generating and power-consuming facilities, and the liberalization of the market demand a reorganization of the decades-old structure of the European electric power supply network into a transport grid for large amounts of commercial power. A study carried out by the German energy agency (dena) in 2010, the dena Grid Study II, investigated the consequences of increasing the wind power generating capacity to 37 GW onshore and 14 GW offshore by 2020, complemented by over 34 GW from photovoltaics, biofuels and geothermal heat. Furthermore, a remainder of 6.7 GW from nuclear plants was assumed, 1.4 GW less than planned in the exit scenario of the federal government in May 2011. The extension of transmission lines by up to 3,600 km and the necessary modifications of the existing lines proposed by this study would lead to costs of up to 1.62 billion € per year. In addition, establishing connections to offshore wind parks would require undersea cables of 1,550 km length, which would lead to further costs of 340 million € per year up to 2020. Financing these extensions of the power grid would lead to price increases of at most 0.5 € cent/kWh. Thus, there are no essential technical hurdles, and the additional costs remain moderate [8].

Since the year 2003, for new installations in regions with major wind power resources, a power generating management has been applied which permits the operators of the transmission grid to reduce or switch off individual power sources when the grid load is too low or when transmission bottlenecks occur. For conventional power plants, this practice leads to savings in the cost of fuel and operations. For wind-power producers, in contrast, it can give rise to serious losses of revenues, since here, the operating and financing costs remain nearly constant.

New plants also require additional capacity in the power transmission network. But the planning of new aerial lines is hampered by public acceptance problems and protracted authorization procedures. Novel approaches, such as the use of conventional buried cables or new bipolar cable concepts with a high transmission capacity, are being pursued only rather hesitantly by the power industry. However, there are still considerable capacity reserves in the present transmission network, if the effective thermal power-transmission limits are exploited in periods of cool weather or strong winds. The measurement of weather data could permit transmission of 30 % more power, and with real-time monitoring of the transmission-line temperature, the increase could add up to 100 % [9]. In Germany, monitoring of this type was introduced in 2006, and it has been practiced in some other EU countries for several years.

The present operational management of the power grid by the four German network operators consists mainly of a permanent adaptation of the generated power input to the varying load. Power generation and purchases are planned 24 hours in advance. By switching on and off of power plants with different regulation time constants, and by short-term buffering using the rotational energy of the generators and turbines, equilibrium is maintained. While up to now, only the load variations and possible power-plant malfunctions had to be compensated, in the future the regulation of the network will be complicated by the variability of the input from wind energy, which has a preferred acceptance status. Wind energy forecast programs are being employed in order to minimize the required capacity of conventional power plants and of additional power reserves. At present, the average deviation of the 24-hour predictions is about 6.5 % (expressed as the mean square error normalized to the installed power capacity) [10].

Considerable deviations in the forecasts can occur in particular due to time offsets in the passage of weather fronts and the corresponding significant power gradients. Under such unfavorable conditions, the input of wind power within a regulation zone can decrease by up to 1 GW per hour and by several gigawatts within a few hours. Further improvements of the forecasts and a reduction of reserve capacity would be possible by using new communications technology, by introducing a more flexible power-plant planning, and by short-term balancing among the different network operators. Reasonable measures include a short-term correction of the 24-hour forecasts, real-time measurements of the output power of wind generators, and the introduction of shorter trading periods on the power market (intraday trading). The earlier dena-I study found that up to the year 2015, no additional power plant reserves will be required to furnish power for regulation and reserves. Furthermore, on the average an hourly and minute-by-minute reserve of conventional power-plant capacity amounting to 8 to 9 % of the installed wind-energy capacity should suffice.

In order to maintain the traditionally very good network stability and supply security in Germany, new grid-connection rules for wind energy generators were introduced in 2003; these require the plants to meet certain criteria. Older, previously installed wind energy plants which correspond to the earlier criteria have to be shut down immediately if network malfunctions occur. This could, in unfavorable cases, lead to a sudden deficit of several gigawatts of input power and produce instabilities in the European electric power network. These risks can however be minimized by modern wind energy plants with transverter technology, by retrofitting of older installations, and by modernization of the power transmission network, which is in any case necessary. Network stability and security can thus be guaranteed even with further increases in the proportion of wind energy.

An increasing proportion of wind energy input power, with its quasi day-to-day variability, will in the medium term

FIG. 7 | PLANT COSTS

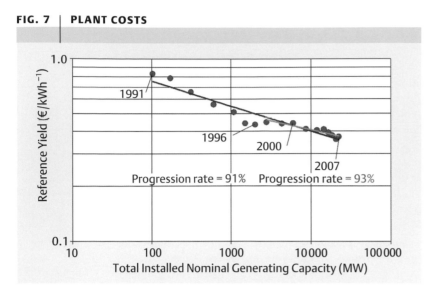

Time evolution of the installation costs relative to the annual energy yield at a reference site, as a function of the total installed generating capacity (graphics: ISET).

require energy storage systems on the scale of power plants. The construction of new pump-storage hydroelectric plants in Germany is not to be expected in the future. Storage via electrolytically-produced hydrogen as an alternative has a very low system efficiency. In the foreseeable future, it will be more reasonable to save fossil fuel by making use of wind energy, and to tide over wind variations by using conventional power plants [11]. Underground adiabatic high-pressure air storage systems, which can yield efficiencies of up to 70 % for thermal energy retrieval, have relatively good future prospects. The methanization of CO_2 would permit the use of the existing natural-gas infrastructure for storage. However, the industrial-scale application of these completely new technologies cannot be expected in the near future. Likewise, the use of decommissioned mine shafts as pump-storage systems is also an interesting option. In the long term, if electric automobiles are in use on a large scale, their energy-storage batteries could be used to equalize fluctuations from wind and solar energy, and thus to stabilize the power grid.

Economic Feasibility

Increasing the use of wind energy in Germany has been stimulated to a major extent by the introduction of a minimum price for wind power and the accompanying planning reliability through the Power Feed-In Law (1991 to 2000) and the Renewable Energies Act (EEG, since April 2000). Thanks to further technical developments and to economies of scale, the costs of wind power plants have been considerably reduced. At present, an installation with 2 MW of output power, 90 m rotor diameter and a hub height of 100 m costs about 2.4 million € ex works, with additional infrastructure costs of 25–30 % at the wind park. At a reference site near the coast (with an annual average wind speed of 5.5 m/s at an altitude of 30 m), about 6.1 GWh per

year can be generated. This is sufficient to supply 1,750 four-person households.

More important than the pure investment costs are the specific generating costs per kilowatt hour produced. Figure 7 shows an inflation-corrected reduction of the plant costs per kWh produced annually at a reference site by over 50 % between 1990 and 2007. This development corresponds to a learning curve with a progression rate of 91 %, increasing to 93 % after 1997. That is, for each doubling of installed power output, the costs fell by 9 % (7 %) (Figure 7).

While in 1991, the subsidized input compensation amounted to a maximum of 18.31 €-ct/kWh, by the year 2006 it had been reduced by 59 % to an average value of 7.44 €-ct/kWh. This historic development is extrapolated in the current Renewable Energies Act (EEG), and is regularly reappraised. The minimum compensation for newly-commissioned onshore plants decreases by 1.5 % per year. Taking inflation into account, new plants therefore have to be more cost effective by about 3.5 %.

Between 2006 and 2008, however, due to the increasing prices of raw materials such as copper and steel and the worldwide increase in demand, the selling price of wind-power plants in Germany increased by nearly 30 %. In the amendment of the EEG which took effect on on Jan. 1st, 2009, this resulted in a slight increase in the compensation. In the later amendment, effective on Jan. 1st, 2012, the compensation was again decreased: Onshore, it was decreased from 5.02 €-ct./kWh to 4.87 €-ct./kWh, and the initial compensation for new plants fell from 9.2 €-ct./kWh to 8.93 €-ct./kWh, while the regression rate was raised from 1 % to 1.5 % per year. The system service bonus decreased from 0.5 €-ct./kWh to 0.48 €-ct./kWh. In 2009, this bonus was first introduced for modern plants, which can improve the stability of the power grid. Offshore, the base compensation of 3.5 €-ct./kWh and the initial compensation of 15 €-ct./kWh remained constant. While there is no regression between 2012 and 2017, a regression of 7 % is to start in 2018. For the so-called repowering (replacement of older plants by new ones with a higher yield), the initial compensation remained at 0.5 €-ct./kWh.

Another provision of the EEG takes into account the importance of the local wind conditions for economic operation of the plants. This determines the amount and the stepwise regression of the different compensation levels during the planned 20-year lifespan of the subsidies. Projects which are obviously economically ineffective have in the meantime been excluded from subsidies. On the other hand, especially favorable conditions apply to offshore sites and to repowering.

The strong worldwide demand for wind-power plants is being driven not only by considerations of environmental and nature protection, but also by the economic rewards of wind power at favorable onshore sites which can be expected in the meantime, in comparison to new construction of conventional power plants. An up-to-date international cost comparison (Figure 8), taking price increases in the power-plant and wind-power markets into account, illustrates this fact.

Nature Conservation and Public Acceptance

With increasing industrial exploitation in the form of large wind parks, wind power has experienced growing acceptance problems. Yet, in comparison to other interventions into nature, such as the increasing concentration of CO_2 and pollutants in the atmosphere, air and ground traffic, aerial transmission lines and many others, wind power installations have however only local and minor negative effects. In view of the directly perceptible consequences of traditional energy supplies, a clear majority of German citizens are still in favor of the continuing development of wind power. Nevertheless, a paradoxical behavior is often observed, accurately characterized by the NIMBY phenomenon, "Not in my back yard!"; i.e. wind power *yes*, but somewhere else.

Thus, for specific wind park projects, a socially and environmentally consistent planning is essential. It has to take into account the interests of the local population as well as recognized minimum standards for nature and landscape conservation. Such an approach can avoid political prejudices and polarization on all sides, that are all too often observed and that cannot be readily countered simply by citing scientific facts or technical solutions.

Ecological and Economical Expediency

With the climate catastrophe looming in the background, German electricity producers are faced with a dilemma. In the coming decades, a major portion of the power-generating capacity must be renewed. At the present level of ca. 544 g of CO_2 emitted per kWh of power generated (2010), Germany lies notably above the European average [12, 13]. A continuation of the current mix of energy inputs with only moderate increases in the proportion of energy from sustainable sources is not a promising alternative, in particular with regard to the self-imposed promise of the German Federal government to reduce CO_2 emissions by 40 % relative to 1990. Also, especially the exit scenario for nuclear power increases pressure to lower emissions by accelerating the development of sustainable energy sources.

Following the decision to adopt this exit scenario, the remaining options for power generation from fossil energy sources are not convincing: On the one hand, the exploitation of the still rich coal reserves in combination with technically immature and economically questionable CO_2 capture and sequestration (CCS), with expected high infrastructure costs and up to 40 % reduction in efficiency [14]; or on the other, a politically risky, in the medium term expensive, and only palliative CO_2 reduction by power generation based on imported natural gas.

In addition to rising fuel costs, the continued use of fossil fuels would be accompanied by ecological and political costs, which arise firstly from the cost of avoiding and overcoming environmental damage, and secondly from one-

FIG. 8 | POWER-GENERATING COSTS

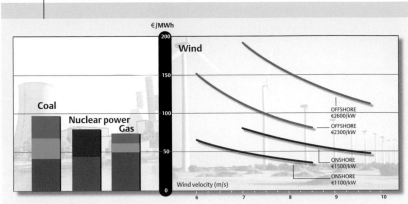

A comparison of the power-generating costs for on- and offshore wind parks with those of various conventional energy carriers. The costs quoted at the lower right are installation costs. In the three bar graphs at the left, the colors refer to the cost of greenhouse gases emitted (incl. fuel production costs) according to the Stern Review (dark green bars represent the upper limits of the estimates), the cost of emission certificates for CO_2 (European market, light green), the spread in costs depending on mining site, ore quality, and fuel treatment (red), and the base costs (purple) (graphics: Windpower Monthly 1/2008).

sided dependence on fuel deliveries from politically often unstable regions. In the short term, such measures might well contribute to a reduction in CO_2 emissions; but in the long run, they are not sound in terms of a sustainable energy supply with predictable, acceptable future costs.

For the renewable energy sources, new challenges present themselves, for example the integration of renewable sources into the power grid, as described above for the case of wind energy, and the compatibility of energy-economy structures. The Institute for Solar Energy Supply Technology (ISET) in Kassel demonstrated in 2005 how the electrical energy supply of Europe and its neighbors could be secured using exclusively sustainable energy sources and currently available technologies, at prices very close to those presently in effect [15]. The central element of such a concept, with a very high proportion of wind energy, is the balancing of input variations from the renewable energy sources among each other. This can be achieved by using a combination of different energy sources and by transporting electric power through a transcontinental power network based on high-voltage direct current transmission (HVDC) with low losses (10–15 %). A similar idea within a smaller framework is the concept of decentralized renewable combined power stations, in which weather and load prognoses serve as the basis for controlling the plants; these are then adjusted to match the real supply and demand for power. Biogas plants and pumped-storage systems could even out the load fluctuations from variations in wind and solar power availability. Initial experience with a pilot plant appears to be promising [16]. It supports and complements the ISET study.

In the framework of an international energy system, the technical and economic perspectives are clearly improved. Assuming further increases in fossil fuel prices, it is predicted that an energy mix from sustainable sources (without photovoltaics) would be more favorable in terms of costs than fossil-fuel power generation by as early as 2015 [17]. The continued rapid deployment of sustainable energies would thus offer a guarantee for stabilizing electric power costs, and thus in the end for the competitiveness of German industry.

Summary

The rapid increase in the utilization of wind energy within the past twenty years was made possible to a large extent by technological developments and a favorable political climate. Alongside the continued improvement of efficiency and economic competitiveness of the wind energy systems, political aspects are now becoming more important. Among these are inte-gration into the national and international power grid and into the international energy economy, as well as a societal consensus concerning energy policy. Power generation from wind energy is thus in transition from an alternative to a mainstream energy source. It can make a decisive contribution in the future to a climate-compatible and economically feasible power generation system.

References and Links

[1] J.P. Molly, Status der Windenergienutzung in Deutschland, Stand 31.12.2011 (DEWI report on wind-energy use in Germany, **2011**); download from: www.dewi.de/dewi/fileadmin/pdf/publications/ Statistics%20Pressemitteilungen/Statistik_2011_Folien.pdf (in German); English site: www.dewi.de/dewi/index.php?id=131&L=0.

[2] R. Gasch and J. Twele (Eds.), *Wind Power Plants: Fundamentals, Design, Construction and Operation*, 2nd Ed., Springer, Berlin, Heidelberg **2012**.

[3] H. Dörner: *Drei Welten – ein Leben, Prof. Dr. Ulrich Hütter – Flugzeugkonstrukteur, Windkraft-Pionier, Professor an der Universität Stuttgart*, 3rd Ed., Windreich, Wolfschlugen, **2009**.

[4] VDMA, BWE: *Windenergie in Deutschland – Inlandsmarkt und Exportgeschäft*, **2011**; download from: www.wind-energie.de/sites/default/files/attachments/press-release/2011/deutsche-windindustrie-maerkte-erholen-sich/windindustrie-deutschland-inlandsmarkt-und-exportgeschaeft.pdf (in German).

[5] WWEA, *World Market recovers and sets a new record*, wwindea.org/ home/index.php?option=com_content&task=view&id=345& Itemid=43.

[6] German Federal Ministry for the Environment, Nature Conservation and Nuclear Safety, National renewable energy action plan. **2010**, www.erneuerbare-energien.de/files/pdfs/allgemein/application/ pdf/nationaler_aktionsplan_ee.pdf (in German).

[7] European Wind Energy Association (EWEA), *Pure Power – Wind Energy Targets from 2020 and 2030*, **2011**, see: www.ewea.org/ fileadmin/ewea_documents/documents/publications/reports/ EWEA_Annual_report_2010.pdf .

[8] German Energy Agency (dena), dena Grid Study II – Integration of Renewable Energy Sources in the German Power Supply System from 2015 – 2020 with an Outlook to 2025, **2010**; see: bit.ly/NYgCh.

[9] Bundesverband Windenergie, Press release on 18.09.2006, www.wind-energie.de .

[10] B. Lange: *Wind Power Prediction in Germany – Recent Advances and Future Challenges*, European Wind Energy Conference (EWEC), Athens **2006**.

[11] D. Stolten: Chapter "Hydrogen: An Alternative to Fossil Fuels?", this book.

[12] German Federal Environment Agency, Climate Change 01/07: *Entwicklung der spezifischen Kohlendioxid-Emissionen des deutschen Strommix*, Berlin, April **2007**.

[13] German Federal Environment Agency, Climate Change 06/03: *Anforderungen an die zukünftige Energieversorgung*, Berlin, Aug. **2003**.

[14] German Federal Office of the Environment, CCS: *Rahmenbedingungen des Umweltschutzes für eine sich entwickelende Technik*, May **2009**.

[15] G. Czisch, *Szenarien zur zukünftigen Stromversorgung – Kostenoptimierte Variationen zur Versorgung Europas und seiner Nachbarn mit Strom aus erneuerbaren Energien*, University of Kassel, Dissertation, **2005**.

[16] H. Emanuel, R. Mackensen, K. Rohrig: *Das regenerative Kombikraftwerk*, Final Report, Kassel, April **2008**.

[17] J. Nitsch, Lead Study **2008**: Study on Development Strategy for Renewable Energies, commissioned by the BMU, Oct. 2008; see: www.bmu.de/english/renewable_energy/downloads/doc/42726.php.

About the Authors

Martin Kühn, born in 1963, studied physics engineering in Hannover, Berlin, and Delft. Until 1999, he was a research assistant at the TU Delft, then through 2003 Project Manager for Offshore Engineering at GE Wind Energy GmbH. In 2001, he completed his dissertation at the TU Delft, and from 2004 to 2010, he held the chair for Wind Energy at the University of Stuttgart. Since 2010, he has been professor for wind energy systems at the University of Oldenburg.

Tobias Klaus, born in 1967, studied political science in Bonn, Frankfurt and Dublin. He heads the development cooperation group at the International Solar Energy Research Center (ISC) in Constance, and works on the interactions between technology and society in sustainable energy projects within development cooperation.

Address:

Prof. Dr. Martin Kühn,
Arbeitsgruppe Windenergie, Institut für Physik,
Universität Oldenburg,
Marie-Curie-Str. 1, 26129 Oldenburg, Germany.
martin.kuehn@uni-oldenburg.de

Tobias Klaus
International Solar Energy Research Center Konstanz
Rudolf-Diesel-Str. 15
78467 Konstanz, Germany
tobias.klaus@isc-konstanz.de

Flowing Energy

BY ROLAND WENGENMAYR

Hydroelectric plants generate nearly one-sixth of the electric power produced worldwide. Water power, along with the biomass, is thus the only sustainable energy source that contributes at present on a large scale to the electrical energy supply for the world's population. It is efficient, but it can also destroy whole regions, societies and ecological systems.

Since the late 19th century, water power has been used to generate electricity. In the year 2009, according to the International Energy Agency, IEA, it provided 16.2 % of the world's total electrical energy requirements, and thus outperformed nuclear power, at 13.4 % [1]. Its contribution to the world's consumption of primary energy, which also includes heat energy, was 2.3 % in the year 2009 [2]. It is thus the only sustainable energy source which presently contributes an appreciable portion of the electrical energy supply for the world's population.

Modern hydroelectric plants achieve a very high efficiency. Up to 90 % of the kinetic energy of the flowing water can be converted to electric power by modern turbines and generators. In comparison, light-water reactors convert only about 35 % of the nuclear energy into electrical energy, while the remainder is lost as "waste heat"; coal-fired power plants have an efficiency of over 40 %, while a modern natural gas-combi power plant achieves over 60 percent.

River and Storage Hydroelectric Plants

The hydroelectric power primer tells us that there are river power plants, storage power plants and pumped-storage power plants. River power plants are used as a rule to supply the base load to the power grid. Their power production depends on the water level in the river, and this varies only slowly over the seasons in most rivers. Storage power plants are usually located high in mountainous regions and collect the water from melting snow in their reservoirs; they are therefore strongly dependent on this seasonal water supply. On the other hand, they can be started up within minutes and can be used to level out variations in grid power. They are thus well suited for compensating peak demand.

Pumped storage power plants, in contrast, are pure energy storage facilities which do not contribute to the production of electrical energy as such. In periods of low demand, they pump water into a high-level reservoir using power from the grid. As required at times of peak demand, they convert the stored potential energy back into electrical energy, by letting the stored water flow back down into the lower storage reservoir – or into a river. They are usually outfitted with special turbines which can work in the reverse direction as pumps. These plants are often used by the power companies for power 'upgrading': they are turned on when power is in short supply and therefore expensive.

Pumped storage plants are at present the only intermediate storage facilities for large amounts of electrical energy. Alternative storage methods, such as air pressure vessels

FIG. 1 | TYPES OF TURBINES

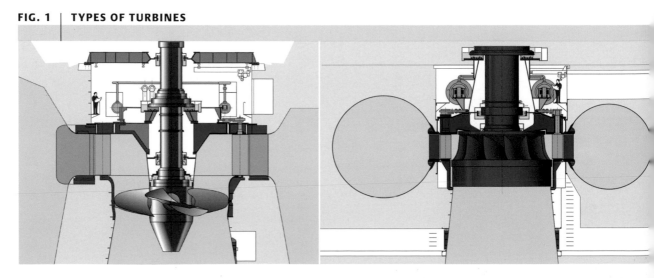

Renewable Energy. Edited by R. Wengenmayr, Th. Bührke. Copyright © 2013 WILEY-VCH Verlag GmbH & Co. KGaA, Weinheim

or the conversion of electrical energy into chemical energy-storage media such as hydrogen or methane, as well as other approaches, are not yet technically mature. As the proportion of fluctuating sustainable energy sources in the grid increases, the need for pumped storage plants will grow. However, this is in opposition to landscape protection interests. An alternative is provided by using decommissioned mine shafts as underground pumped storage plants, for example in the Ruhr region of Germany.

The most modern pumped storage power plant in Germany went online in 2003 in Goldisthal, Thuringia. Its turbines were supplied by *Voith Hydro*. This company, located in Heidenheim, is one of the world's major producers of equipment for hydroelectric power generation. As its technical board member Siegbert Etter explains, no other machines attain a power density as high as that of modern water turbines. A turbine which can deliver a hundred kilowatts of power is only 20 cm (8") in diameter, and is thus much more compact than an automobile engine of similar power. The amount of power which a turbine can deliver depends essentialy on the velocity and the amount of water that flows through it per unit time.

Large-scale Hydroelectric Plants

A typical flow power plant in a river without a significant head of water accepts a large volume of water at a relatively sluggish velocity. It uses Kaplan turbines, which are reminiscent of enormous ship's propellers (Fig. 1, left side). At low rotational speeds, they extract the optimal amount of useful energy from the low water head at typically moderate flow velocities. The plant operators can adjust the pitch of the turbine blades and the fin-like guides in the housing through which the water flows into the turbine.

As the water head becomes higher, its kinetic energy also increases. Power plant owners therefore take advantage of the differences in altitude in mountainous regions, where storage reservoirs collect melt water at high levels. It flows down hundreds of meters through shafts and pipes to the power plant, where it jets out of nozzles into the massive

The Three-Gorge dam in China during construction in 2003. Behind its walls, which are up to over 180 m high, the water of the Yangtze River has meanwhile backed up to form a lake more than 600 km long (photo: Voith Hydro).

buckets of Pelton turbines (Fig. 1, right side). These modern descendents of the water wheel have dividing partitions in the centers of their buckets, which split the water jets as they hit the turbine. The curved buckets deflect the water jets by nearly 180°, causing a maximum change in the momentum of the water and allowing the turbine wheel to extract nearly all of its kinetic energy. With a head of 1000 m, the water bursts out of the nozzles at up to 500 km/h and drives the turbines up to 1000 rpm. The nozzle openings can be adjusted by cones and a pivoted flow deflector directs the flow of water to the turbines.

The largest power plants are built along the earth's greatest rivers (Figure 2). Their massive dams do not produce a very high water head, but their turbines handle extremely large volumes of water. The controversial Three Gorge Project in China is currently the largest hydroelectric plant in the world. Its dam is over 180 m (nearly 600) high and has at present backed up the Yangtze River to form a lake 660 km (410 mi.) long. The 26 giant turbines deliver nominally 18.2 GW of electric power. This corresponds to 14 nuclear power plant blocks or 22 large coal-

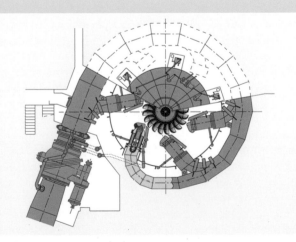

Left: The water (blue) flows horizontally into this vertical-shaft Kaplan turbine past the control vanes (green).
Center: Francis turbines are usually mounted with their shafts vertical. The water flows in radially past the control vanes (green) and exits axially down the "outlet pipe".
Right: A Pelton turbine (red) with an input pipe (penstock) leading to six steerable nozzles (one shown in cross-section); to the right of each nozzle is a flow deflector. A portion of the housing with the mechanical controls is indicated schematically (graphics: Voith Hydro).

FIG. 2 | A RIVER POWER PLANT

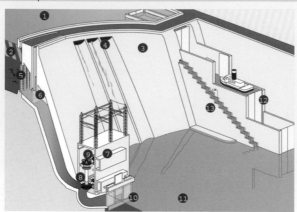

In a large-scale river power plant, the water at the upper level of the river (1) fills a reservoir. In order to allow controlled runoff of high water (higher water level (2)), the dam (3) has a spillway (4). In normal operation, the water flows past grids which can be raised to catch flotsam (5) and sluice gates (6) through "penstocks" down to the powerhouse (7) and the turbines (8). The vertical-shaft Francis turbines drive the generators (9) via connecting shafts. The water then exits through lower sluice gates (10) to the lower river level (11). The sluice gates (6) and (10) are water-tight when closed, so that each turbine can be emptied of water and inspected. A system of locks (12) allows ships to pass the dam. The fish ladder (13) attracts the fish with a current of water and encourages them to choose this safe route (graphics: Esjottes/von-Rotwein, Illustration + Infografik).

fired plants [3]. The power actually obtained in practice is however around half of this value. In 2008, this plant produced over 80 TWh of electrical energy, which would cover about 13 % of the electrical power requirements in Germany. Its Francis turbines were designed by *Voith Hydro*. Each of them is 10 meters in diameter and weighs 420 tons.

Francis turbines can accept a large range of water velocities corresponding to moderate up to very high water heads (the latter are are the domain of the Pelton turbines). Their curved buckets are not adjustable (Fig. 1, center). The water flows through a delivery pipe (spiral) radially into the turbine and causes it to rotate. It then exits downwards along the turbine shaft through the outlet pipe to the lower water level of the river. Regulation is provided by the control surfaces arranged around the perimeter of the turbine wheel, whose jets are adjustable. The giant Francis turbines of the Three Gorge project operate at 75 rpm. With a water head of 80 meters, the massive water columns of the Yangtze flow into the turbines at a velocity of 20 km/h. The water is accelerated up to 120 km/h on the rotating turbine buckets.

Modern Francis turbines extract almost all of the kinetic energy from the water and produce a large drop in pressure. This reduces the pressure at the outlet so drastically that the water foams up in cold water-vapor bubbles. This 'cavitation' has to be taken into account by the engineers when designing the turbines, since it must never be allowed to come into contact with the turbine blades. If the bubbles touch the metal, they implode violently and produce cavities in the surface of the blades (thus the term 'cavitation'). The turbines must be constructed and operated in such a way that cavitation occurs only at their outlets.

Each power plant has its own unique characteristics, and water turbines are tailor-made for a particular plant. The firm in Heidenheim currently designs them using complex computer simulations, and optimizes the design using small-scale models in their own test bed. This is itself a small power plant with a megawatt of output power. Ecological considerations can also influence the design of the turbines. In the USA, some power plants employ turbines which blow air into the water through special channels, thus increasing the low oxygen content of the river. There are even 'fish friendly' turbines: Fish which have missed the fish ladder (Figure 2) have a chance of survival when passing through them.

Small-scale Hydroelectric Plants

Fish can also be an issue even for the constructors of small power plants. In order not to endanger the fish population of the small Black Forest River Elz, the water-power equipment company *WKV* in Gutach constructed an elaborate inlet structure. A fish ladder and a fine grid keep the fish from entering the kilometer-long pipe which carries water parallel to the river into the turbines of the heavy machine factory. The factory obtains its electrical power to a large extent from the water of the Elz. The two Francis turbines with a total power output of 320 kW produce more power in the course of a year than *WKV* needs for its production lines. They deliver the excess power to neighboring households.

WVK supplies the market for small and medium-sized water-power plants. The firm was founded in 1979 by a teacher, Manfred Volk, as a 'garage operation'. It has been growing ever since, and has delivered plants to over 30 countries. Its turbine technology is developed by *WKV* in cooperation with the Technical University in Munich.

This Breisgau firm is successful, but it has to deal with the vagaries of the sustainable-energy market. According to *WKV*'s financial director Thomas Bub, 70 to 80 percent of the projects planned by potential customers fizzle out for lack of financial backing. The particular economic aspect of water power lies in its extremely long useful life: Some plants are in use for more than 80 years. They can take full advantage of the cost-free energy supply over this long time period. On the other hand, the initial investment is often considerably higher and more complex than for a comparable fossil-fuel power plant. The interest payments on the high capital investment at the outset mean that many plants are amortized only after several decades. Thus, hydroelectric power needs investors and creditors who think in long terms.

Large Dams and their Consequences

This is particularly true of the billion-dollar investments required for large-scale hydroelectric plants. Their negative image has dampened willingness to finance the investments on the part of traditional major backers such as the World Bank. Hydroelectric power on a large scale always exacts a high price. It can damage or even destroy whole regions, ecosystems and social structures. The World Commission on Dams (WCD) listed 45,000 large dams in the year 2000 [3]. Half of these dams were constructed for power production.

The WCD estimates that for the construction of these dams, worldwide 40 to 80 million persons were displaced or forced out of their homes [3]. A famous negative example is the 50-year-old Kariba dam in Zimbabwe. It caused massive changes in the delta of the Zambezi River. 60,000 persons were forced to move by the construction of the reservoir [3]. In the case of the Chinese Three Gorge project, apparently more than a million people had to be relocated.

This policy has generated massive criticism on the part of non-governmental organizations such as the International Rivers Network. They have in the past applied political pressure on the World Bank so successfully that it practically withdrew from financing large hydroelectric projects in poorer countries. However, new financial sources have appeared which have again stirred up activity in this area. In particular, India and China have offered partnerships to countries which are poor in capital but rich in water. They have shown fewer ethical scruples in their financing agreements and offer to deliver the technology at favorable prices at the same time. This has forced the World Bank to make another policy reversal, to avoid avoid being left out of the process altogether [4].

The safety of dams is another problem which must be taken seriously. In 1975, the Banquiao dam in China burst after a typhoon which was accompanied by catastrophic rainfall, causing 26,000 fatalities. This was the greatest man-made catastrophe in history, says Stefan Hirschberg of the Paul Scherrer Institut in Villigen, Switzerland. But he also points out that in the OECD countries, there has been only a single accident since 1969, which occurred in the USA and caused only a few deaths.

Hirschberg carries out systems research on energy technologies and is well acquainted with the situation in China. From his point of view, there are many advantages not only to small-scale but also to large-scale hydroelectric power. A major plus is the (often) low emission rate of greenhouse gases. Reservoirs can – depending on their geological and climatic situation – emit carbon dioxide and, as a result of the decomposition of plant material, also methane.

But even when the production of the materials for constructing a typical hydroelectric plant is taken into account, on the average only the equivalent of a few grams of CO_2 per kilowatt hour are released, Hirschberg explains. A typical coal-fired power plant, in contrast, emits a kilogram of CO_2 per kilowatt hour of energy produced, and a Chinese coal-fired power plant emits up to 1.5 kg per kWh.

The outdated Chinese power plants also lack smoke filtering systems. In densely-populated areas, they shorten the lifespan of the population measurably. According to Hirschberg's studies, 25,000 years of life expectancy are lost there per gigawatt of power generated each year. Furthermore, the coal mines degrade the overall ecological and social conditions in China. They emit enormous amounts of methane, and between 1994 and 1999 alone, more than 11,000 miners lost their lives in accidents [5].

These statistics for a country with 25 % of the world's population make it clear that large-scale hydroelectric power plants may be the lesser evil. Hirschberg in any case maintains, "In the context of global climate policy, hydroelectric power occupies an excellent position".

Summary

Water power generates nearly one-sixth of the electric power produced worldwide and thus greatly outpaces all the other forms of sustainable energy. Kaplan turbines are suitable for electric power generation with a low head of water, while Francis turbines are used for moderate hydraulic gradients and Pelton turbines for very large gradients with high flow velocities. There are river power plants and storage power plants. Pumped-storage power plants are used purely for energy storage. Modern water turbines and generators can convert the kinetic energy of the water into electrical energy with up to 90 % efficiency. However, large-scale power plants can destroy whole regions, societies and ecosystems.

References

[1] International Energy Agency, Key World energy Statistics **2011**, 24.
[2] cf. [1], p. 6.
[3] "Dams and Development", Final report in the year **2000** of the World Commission on Dams, www.unep.org/dams.
[4] Henry Fountain, "Unloved, but not Unbuilt", The New York Times **2005**, June 5.
[5] S. Hirschberg *et al.*, PSI Report No. 03–04, Paul Scherrer Institute, Villigen **2003**.

About the Author

Roland Wengenmayr is the editor of the German physics magazine "Physik in unserer Zeit" and is a science journalist.

Address:
Roland Wengenmayr, Physik in unserer Zeit, Konrad-Glatt-Str. 17, 65929 Frankfurt, Germany. Roland@roland-wengenmayr.de

Solar Thermal Power Plants

How the Sun gets into the Power Plant

BY ROBERT PITZ-PAAL

Large, precisely curved mirror surfaces and enormous heat storage tanks are the most obvious components of solar-thermal power plants, which are rapidly multiplying in the desert regions of the world. They collect the sunlight and concentrate it onto a thermal power-generating unit.

Today, when people talk about solar power, they usually mean power produced by the shiny blue photovoltaic cells on the roofs of houses or along expressways. It is practically unknown that solar thermal power plants, which are based on a completely different operating principle, already feed more than 3 TWh per year of electric energy into the power grids worldwide – tendency increasing [1]. The origins of this technology were in Europe [2]. There, it is now advancing rapidly, since Spain and Italy have begun subsidizing its commercial exploitation.

Germany, Austria and Switzerland are too far north to be able to operate solar thermal power plants economically. Nevertheless, German research organizations and firms in particular are contributing intensively to further development of solar thermal power generation technology for the export market. In the future, the import of solar-thermally generated electric power could also become an important factor for the northern industrial countries in helping to reduce their CO_2 emissions [3,4].

The Principle

Nearly 80 percent of the electric power that we use comes from fossil-fuel or nuclear power plants. The principle of power generation is in all cases the same: Heat energy from combustion of fossil fuels or from nuclear fission is used to drive a thermal engine – as a rule using a steam turbine cycle – and to produce electric current via generators coupled to the turbines. Solar thermal power plants use exactly the same technology, which has been refined for more than a hundred years. They simply replace the conventional heat sources by solar energy.

In contrast to fossil energy sources, solar energy is not available around the clock. The gaps, for example at night, can be bridged over in two ways by the power plant operators: Either they switch to fossil fuel combustion when the sun is not available, or else they store the collected heat energy and withdraw this stored heat as needed for power generation.

Solar thermal power plants work in principle like a magnifying glass (Figure 1a–c). They concentrate the rays of the sun, in order to obtain a high temperature: At least 300° C is required in order to be able to generate power effectively and economically with their heat engines using the collected solar energy. The flat or vacuum-tube collectors familiar from rooftop applications are not suitable.

The required high working temperature necessitates strong, direct solar radiation, and this determines which locations are appropriate for solar thermal power plants. They can thus be operated economically only within the – enormous – Sun Belt between the 35th northern and 35th southern latitudes. This distinguishes them from photovoltaic systems, which can generate power effectively even with diffuse daylight, and are therefore also suitable for use under the conditions in Central Europe.

The Concentration of Light

If a black spot is irradiated by sunlight, it will heat up until the thermal losses to its environment just compensate input of solar radiation energy. When useful heat is extracted from the spot, its temperature will decrease. To reach higher temperatures, there are two possibilities, which can be used in parallel: reduction of the thermal losses, and an increase in the radiation energy input per unit area. The latter requires concentration of the direct solar radiation, which can be accomplished by using lenses or mirrors. But how strongly can solar radiation be concentrated?

The solar disc has a finite size; from the earth, its diameter appears to us to correspond to an angle of about one-

Renewable Energy. Edited by R. Wengenmayr, Th. Bührke. Copyright © 2013 WILEY-VCH Verlag GmbH & Co. KGaA, Weinheim

Fig. 1 *(a) Prototype of the improved Euro-Trough parabolic collector* (photo: DLR). *(b) The solar tower power plant CESA 1 at the European test center 'Plataforma Solar de Almería' is currently a test platform for various new developments. (c) The European Dish-Stirling system called 'EuroDish'.*

half of one degree. For this reason, not all the sun's rays which reach the earth are precisely parallel to one another. That would, however, be required in order to be able to concentrate them onto a single point. Therefore, the maximum possible concentration is limited to a factor of about 46,200-fold. Nevertheless, we can in this way theoretically arrive at about the same radiation energy density as on the surface of the sun, and could in principle obtain heat at a temperature of several thousand Kelvins. The focus point of the concentrator has to be at the same place during the whole day; the concentrator thus has to follow the sun by moving around two axes. An alternative is offered by linear concentrators, for example cylindrical lenses: They do not concentrate the radiation at a single point, but rather along a caustic line, so that they need to be moved around only one axis in order to follow the sun. In this case, however, the theoretical maximum degree of concentration is only 215-fold. This is still sufficient to obtain useful heat at a temperature of several hundred Kelvins.

Concentrating Collectors

In practice, mirror concentrators have for the most part taken predominance over lens concentrators [7,8]. They are more suitable for assembly on a large scale and are less costly to construct. Essentially four different structural types can be distinguished (Figure 2). The *dish concentrator* is

an ideal concentrator which follows the motion of the sun along two axes. It consists of a parabolic silvered dish, which focusses the radiation of the sun onto a single point. The receiver for the radiation, and often a heat engine which is directly connected to it, are mounted at the focal point, both fixed in relation to the dish, so that they move with it. Wind forces, which deform the surface of the concentrator, limit its maximum size to a few 100 m^2 and its electric power output to a few tens of kW.

The *central receiver system* solves this problem by dividing up the oversized parabolic concentrator into a set of smaller, individually movable concentrator mirrors. These *heliostats* are directed onto a common focal point at the top of a central tower ('tower power plant'). There, a central receiver collects the heat. Since such a concentrator is no longer an ideal paraboloid, the maximum possible concentration factor decreases to 500- to 1000-fold. This, however, is sufficient to reach temperatures of up to 1500 K. Large central receiver systems with thousands of heliostats, each with 100 m^2 mirror area, would require towers up to 100–200 m high. They could collect several hundred MW of solar radiation power.

The *paraboic trough concentrator* is a linear concentrator which is moved around only one axis. A parabolic silvered trough concentrates the solar radiation up to 100-fold onto a tube which runs along the caustic line, and in

FIG. 2 | CONCENTRATION OF THE SUNLIGHT

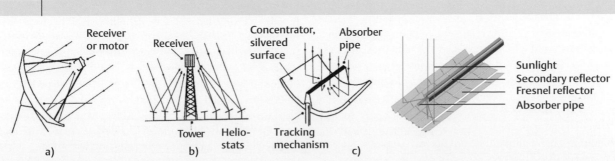

Four different solutions for concentrating the solar radiation: a) dish concentrator, b) central-receiver system, c) parabolic trough, d) linear Fresnel collector.

FIG. 3 | A SOLAR THERMAL POWER PLANT

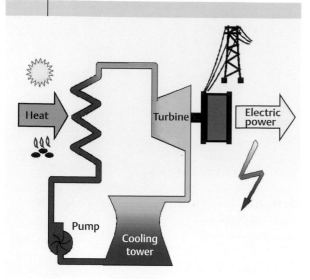

*High-tempera-
ture heat energy
from the sun or
fuel combustion
drives a steam
turbine.*

which a heat-transfer medium is circulated.

The *linear Fresnel collector* is a variant of the parabolic trough concentrator. Its parabolic reflector is sliced into narrow strips which are arranged beside each other like the vanes of a venetian blind, and thus offer little resistance to wind forces. It permits the construction of very large apertures. A further advantage lies in the fact that the absorber tube need not be moved along with the reflector, but instead can be installed in a fixed position, simplifying its connection to the heat-transfer piping. These advantages are however bought at the cost of a reduced optical efficiency, depending on the latitude of the installation. In order to collect the same amount of solar energy, 15 to 40 % more collector area is thus required (Figure 2d). The theoretical maximum value of 215-fold concentration is not attainable in practice for two reasons: On the one hand, the large troughs "lie" on the earth's surface and therefore cannot be rotated around all possible spatial axes to point towards the sun; and on the other, surface imperfections reduce the geometric quality of the mirrors.

Trough collectors can be joined up into line sections many hundreds of meters in length. Numerous parallel lines can then collect hunderds of MW of thermal power for one power plant block.

Heat-Engine Processes

A thermal power process can unfortunately not convert all the heat energy provided to it into mechanical work. It follows from the Second Law of Thermodynamics that part of the heat energy must be extracted from the process at a lower temperature than the input heat (so called "waste heat"). The higher the input temperature and the lower the output temperature, the greater the fraction of heat which can be converted into mechanical work (and thus into electric power in a power plant). It therefore follows from thermodynamics that a high "temperature head" between the hot and the cold heat reservoirs is more favorable for the thermal efficiency than a lower one.

In conventional thermal power processes to which solar energy can be applied, steam thermal engines (Clausius-Rankine process) are very often used: Water is vaporized at high pressure in a boiler and the steam is further superheated. This hot steam expands in a turbine and performs

mechanical work there. It is then condensed back to liquid water in a condenser and flows back into the boiler (Figure 3). The cooling of the conden-ser removes a part of the thermal energy from the process cycle and thereby fulfills the laws of thermodynamics.

Modern steam power plants operate at steam pressures above 200 bar and at temperatures of over 600° C. As a rule, they generate an electric output power of several hundred MW_{el}. As a result, parabolic-trough concentrators and central-receiver systems in particular are suitable sources of heat for this type of power plant, while dish concentrators can be used to drive other, more compact thermal engines with lower power outputs.

The solar energy concentrators in use today cannot quite reach the extreme steam temperatures and pressures mentioned above. The levels they reach nevertheless permit reasonable and efficient power generation, if the steam power plant is designed to suit them. Since central-receiver and dish systems can in principle produce notably higher temperatures, it makes sense to utilize this potential. Due to the higher temperature of the input heat, the power plant can convert more heat into electrical power per unit area of its mirrors. As a result, for the same power output it requires less concentrator area, which saves on costs for its construction and operation.

Gas turbines represent a mature technology as heat-engine power plants, with high operating temperatures of 900 to 1300° C. For simplicity, they make use of air as operating medium. However, the outlet air temperature is still rather high at 400 to 600° C. They thus offer no efficiency advantage in comparison to steam systems; only with a combination of gas and steam turbines ('combined-cycle' or CCGT plants) can the desired improvement be realized. In this concept, the sun preheats the gas-turbine cycle, and the 'waste heat' output from the gas turbines is fed to steam generators for the separate steam-turbine cycle. Such systems can generate 25 to 35 % more electric power from a kWh of heat energy than a pure steam turbine plant.

Small gas turbine systems (without steam turbines) can also be used in dish concentrator systems. To date, however, Stirling engines have been predominant: In contrast to internal-combustion engines, these hot-air engines require an external source of heat, as can be provided optimally by the focal point of a parabolic mirror reflector; on the other hand, they require no fuel. Additional advantages of Stirling engines are their high thermal efficiencies and their hermetically-sealed construction, which reduces maintenance costs. Since the market for Stirling engines has been small up to now, the choice of available models is still relatively limited.

Parabolic-Trough Power Plants

Parabolic-trough power plants were the first type of solar thermal power plants to generate electric power on a commercial basis. As early as 1983, the Israeli firm *LUZ International Limited* closed a contract with the Californian energy supplier *Southern California Edison* (SCE) to deliver power from two parabolic-trough power plants called SEGS (Solar Electricity Generating System) I and II. By 1990, all together nine power plants had been built at three different locations in the Mojave Desert in California, with an overall power output of 354 MW_{el} and more than two million square meters of collector area. In order to be able to deliver power reliably during peak use periods, these power plants are allowed to supply 25 % of their thermal input energy from combustion of natural gas.

However, since fossil fuel prices did not rise as originally expected, but instead fell, it was not possible to build additional power plants cost-effectively. The existing solar power plants continue in service and feed nearly as much electrical power into the grid as all the photovoltaic systems worldwide, as mentioned above.

The passage of a power feed-in law in Spain in 2004 has rewarded electrical energy produced in solar thermal plants and fed into the grid at up to 28 €-cent/kWh. This has led to a veritable boom in plant construction. By mid-2011, in Spain alone there were plants with 730 MW output power in operation and more than 800 MW under construction. Worldwide, the corresponding figures are 1200 MW and 2300 MW. In addition, several GW are in the concrete planning stage. The market is currently concentrated in Spain and the USA, but additional projects are under construction in Algeria, Egypt, Morocco, and the United Arab Emirates as well as in India, China and Australia. While at first, mainly parabolic-trough plants with a power output of 50 MW were constructed, some of them outfitted with heat-storage systems permitting 7 hours of full-power operation without sunlight, in the meantime plants of higher output power (up to 250 MW) and also using other technologies (Fresnel, central tower, dish) have been put into service.

All the commercial parabolic-trough plants make use of a synthetic thermal oil which is heated up to 400° C on passing through the collector and then flows through a steam generator.

The collectors have typical apertures of around 6 m. A single hydraulic drive moves a collector section up to 150 m in length around one axis to follow the sun. The absorber tube in which the heat-transfer medium flows is made of steel and has an optically selective outer coating, which absorbs radiation within the solar spectrum effectively but re-radiates only a small amount of heat and thus minimizes heat losses to the environment. To further reduce losses, the absorber tube is surrounded by an evacuated glass envelope. The mirror segments are made of thick glass with a reduced content of iron, which would absorb the light and is therefore unwanted. The glass is silvered on its rear surface.

Based on the first three generations of collectors in the SEGS plants, various manufacturers worldwide have developed the technology towards increased stiffness, improved optical precision and simpler mounting. This results in increased specific earnings of up to 10 % per collector. For the

key components, such as curved glass mirrors and absorber tubing, there are currently several active manufacturers who have continued to improve their products in terms of efficiency and service life. New developments aim at apertures of up to 7.50 m. Linear Fresnel systems are also commercially available from several manufacturers at present. However, these systems are currently limited to the generation of saturated steam at temperatures below 300 °C.

Central Receiver Systems

Central receiver systems are still in the early phases of commercial operation. Since the beginning of the 1980's, around the world more than ten smaller demonstration plants with central receivers have been put into service (see Table 1 and Figure 1b). Their operation was however terminated after the end of the test campaigns, since they were too small to be operated cost effectively. Only since 2007 have the first commercial plants begun operation, especially in Spain. They have made use initially of relatively moderate steam parameters in order to guarantee safe and low-risk operation (see Table 1). In follow-up projects, it is planned to increase the operating temperatures step by step and thereby to improve their efficiencies [8].

The electric power was generated by a steam turbine in all these test plants. The main difference among the various test plants lies in the choice of the transfer medium used to transport heat energy from the top of the receiver tower to the steam generator. It first appeared attractive to use the steam itself as thermal transfer medium; this would eliminate the need for intermediate heat exchangers or steam generators and allow a direct connection to the steam turbines.

However, this concept soon showed two essential faults. In the first place, it was not easy to control the generation of superheated steam in the receiver under conditions of fluctuating solar radiation input, since the pressure and the temperature of the steam must be kept nearly constant in the turbine circuit. Secondly, with practicable technology it was nearly impossible to store heat energy within steam without considerable thermodynamic losses. In present-day commercial plants, therefore, the use of superheated steam has not yet been implemented. Additional projects are however underway in the USA and Spain which will eliminate this restriction and generate superheated steam.

In a parallel development, the use of molten sodium as heat-transfer medium was tested. After a serious fire at the European test site Plataforma Solar de Almería in southern Spain, however, it became apparent that this highly reactive

FIG. 4 | ABSORBERS

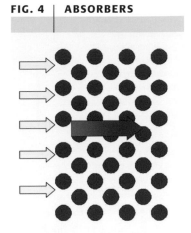

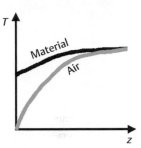

Above: Radiation (yellow arrows) is incident from the left onto a porous structure (red dots), while air passes through the structure from the left to the right; it takes up heat (blue-red arrow). Below: The dependence of the temperatures of the material in the structure and of the air as a function of the depth z within the structure.

FIG. 5 | **DIRECT STEAM GENERATION**

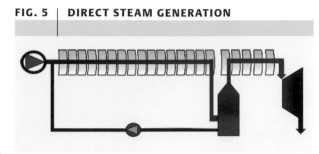

Direct solar steam generation in parabolic troughs: The water is partially vaporized in the first two-thirds of the collector line. Then the mixture of steam and water is separated and the dry steam is further superheated in the last third of the line, while the hot water flows back to the inlet of the collectors.

metal is too dangerous. In the early 1990's in America, the concept of using molten salts as heat transfer medium, which originated in France, was further developed and demonstrated between 1996 and 1999 at the 10-MW plant 'Solar Two' in Barstow, California.

Mixtures of potassium and sodium nitrate salts can be optimized in terms of their melting points to the parameters required for steam generation. They offer two advantages: The relatively low-cost salts have good heat transfer properties; and furthermore, they can be stored at low pressure in tanks for use as a thermal storage medium. This makes it unnecessary to exchange heat with an additional storage medium. Their disadvantage is their relatively high melting points, which, depending on the composition of the mixture, lie between 120° C and 240° C. This necessitates electrical heating of all the piping to avoid freezing out of the salts and resulting pipe blockage, for example during system start-up.

On the basis of experience gained with Solar Two, a larger successor plant is being constructed in Spain. Gemmasolar is expected to attain an output power of 15 MW_e using a mirror area (solar field) increased by a factor of three, and its storage reservoir will be able to store sufficient energy for 16 hours of electric power generation.

The third concept makes use of air as heat-transfer medium. Air has, to be sure, rather poor heat transfer properties, but it promises simple manageability, no upper or lower

limits to the operating temperature, unlimited availability and complete lack of toxicity. Air also conjures up the vision of being able to operate combined gas and steam turbines at a high temperature for the first time using solar energy; these would make more efficient use of the collected solar heat, and thus of the mirror surface area.

In the first test setups, it was attempted to heat the air by irradiating bundles of pipes through which it was passed. But only since the development of the so-called volumetric receiver has it become possible to adequately compensate the poor heat-transfer properties of the air. Such a receiver contains a 'porous' material, for example a meshwork of wire, which is penetrated by the concentrated solar radiation and through which at the same time the air to be heated flows (Figure 4). The large internal surface area guarantees efficient heat transfer. If the air circuit is open and operates at atmospheric pressure, then such a receiver can drive steam turbines. If, on the other hand, the air receiver is closed with a transparent radiation window and the air is pressurized, then the system can even be used with gas turbines.

Air systems at atmospheric pressure are practically free of operating disturbances, and this is the reason why they are favored by a European consortium: In 1994, at the Plataforma Solar, a 3 MW test system operated without problems on the first try. In the meantime, further research conducted by the DLR within the European Network was able to increase the efficiency of individual components and to reduce the costs of the receiver and the storage reservoir. A first demonstration project with 1.5 MW of electric output power is currently being set up at Jülich in Germany, and will serve as a technical benchmark for potential manufacturers.

Dish-Stirling Systems

Dish-Stirling systems are at present the least technologically mature. Companies in the USA and in Germany are currently working on four different systems worldwide (Figure 1c). The system which is furthest along in its development originated in Germany and has accumulated several tens of thousands of hours of operation.

Such systems aim at independent power generation, not coupled to a power grid, for example for providing isolated villages with electric power. Their principal advantage is a very high efficiency of up to 30 %: This is provided by the combination of a nearly ideal paraboloid concentrator with an excellent heat engine. If the sun is not shining, then dish-Stirling systems can in principle be operated with fuel combustion, in order to meet the demand for power. This is a decisive ad-

TAB. 1 | **OVERVIEW OF CENTRAL RECEIVER SYSTEMS WORLDWIDE**

Project Name	Country	Power (MW$_e$)	Heat transfer medium	Heat storage medium	Start of Operation
SSPS	Spain	0.5	Liquid sodium	Sodium	1981
EURELIOS	Italy	1	H$_2$0 steam	Nitrate salts/H$_2$O	1981
SUNSHINE	Japan	1	H$_2$0 steam	Nitrate salts/H$_2$O	1981
Solar One USA	USA	10	H$_2$0 steam	Oil/stone	1982
CESA-1	Spain	1	H$_2$0 steam	Nitrate salts	1983
MSEE/Cat B	USA	1	Nitrate salts	Nitrate salts	1983
THEMIS	France	2.5	Hi-tech salts	Hi-tech salts	1984
SPP-5	Ukraine	5	H$_2$0 steam	H$_2$0 steam	1986
TSA	Spain	1	Air	Ceramic reservoir	1993
Solar Two	USA	10	Nitrate salts	Nitrate salts	1996
PS 10	Spain	11	Saturated steam	Steam	2007
Solar Tower	Jülich (Germany)	1.5	Air	Ceramic reservoir	2008
PS 20	Spain	20	Saturated steam	Water/Steam	2009
Gemmasolar	Spain	19.9	Nitrate salts	Nitrate salts	2011

vantage over photovoltaic cells, which aim at a similar market: They, however, require expensive storage batteries for energy storage.

These are good reasons why dish-Stirling systems have a favorable market chance in the medium term for independent power generating applications. For this purpose, they must be capable of autonomous and very reliable operation. Subsidized niche markets are, however, only one of the possibilities for dish-Stirling systems. A still greater market potential lies in the increasing power requirements of developing countries, especially those with a large amount of sunlight, poorly established power grids and high costs for the import and transport of fossil fuels.

Aside from the question of technical maturity, the small number of units produced represents a hurdle to commercial marketing of dish-Stirling systems.

Cost Effectiveness

In the research and demonstration systems of the 1980's, the costs of power generation were still in the range of 50 to 100 €-cent/kWh. The SEGS power plants were the first to reduce these costs significantly with their commercial technology. In the first SEGS plants, they were around 30 €-ct./kWh; with technical improvements and upgrading, they sank to about 20 €-ct./kWh. Today's solar plants have offered prices as low as 14 €-ct./kWh.

The profitability of a solar thermal power plant naturally depends strongly on its location. The available solar energy influences the costs per kWh approximately linearly. At the SEGS sites in the Mojave Desert in California, annually about 2.5 times as much direct solar radiation is available as in Germany, and 25 % more even than in Southern Spain.

If one assumes the same conditions of insolation and compares them with good sites for wind power plants, the result is that electric power from solar thermal power plants is at present about two to three times more expensive than power from wind plants and slightly higher than power from photovoltaic cells. However, power from solar thermal plants can be provided much more flexibly to the grid using low-cost thermal energy storage and therefore represents a significantly higher value as grid input [10,13]. In computing the costs, one must distinguish between large installations with several tens of MW_e of electrical output power, and small applications which are not connected to the power grid. The numbers quoted above hold for large plants and still contain considerable potential for cost reduction.

Solar thermal power plants can store their energy in intermediate thermal storage reservoirs at low cost and sell it as needed. They thus represent a fully-fledged replacement for conventional power plants, however with no CO_2 emissions. Increasing fuel prices and CO_2 penalties lead us to expect a cost increase for new conventional power plants of up to 8 to 10 €-ct./kWh – a value which solar thermal power plants could attain or even undercut within the next 10 years [9,10]. It thus appears quite reasonable that the marketing of these technologies should be promoted by various organizations and agencies. Beyond the use of solar thermal power plants for the local supply, also the export of solar power from the deserts of North Africa to Central Europe is currently being discussed in terms of the DESERTEC project (see also chapter "Power from the Desert"). 15 % of flexible power from North Africa would be sufficient to offset fluctuations from photovoltaic and wind power in the European grid by using power from stable and flexible solar thermal plants [12]. However, new grid connections between Europe and North Africa will be necessary for this option.

Technical Improvements

A clear-cut cost reduction can be expected from the following factors: automated mass production of a large quantity of components; an increase in reliability of the plants; and extensive automation of plant operation as well as of the cleaning of the collectors. An important contribution is also promised by further improvements in the technology and innovative concepts for large solar-thermal plants. In Germany, these goals are being pursued at the German Aerospace Center (DLR) as part of its energy research program, together with industrial partners. We will mention briefly some of these research activities here.

An important aspect is increasing the operating temperatures, which, as explained above, will improve conversion efficiencies and permit a smaller specific collector area to be used. For parabolic-trough collectors, the operating temperature limit of the thermal oils used must be increased to above 400° C. One possibility, which has already been tested, is the direct evaporation and superheating of water in the collector itself (Figure 5). For this test, a 500 m long collector loop was set up at the Plataforma Solar in Almería. Among other things, the regulation behavior and flow properties of the water-steam mixture in absorber tubes is investigated with this apparatus. More than 10,000 hours of test operation have proven the technical feasibility of this concept. It should yield a decrease in the power-generating costs of around 10 %. A prototype plant with an output capacity of 5 MW was put into operation in Thailand in 2011. The use of molten salts, similar to the concepts described above, is also an option for increasing operating temperatures; this has been demonstrated on the 5 MW scale in Italy. Here, a particular challenge is to avoid freezing out of the salts in the more than 100 km long piping systems of full-sized commercial plants.

Intensive research is being carried out on central-receiver systems with the goal of using pressurized air as the heat-transport medium for solar energy, allowing a high input temperature for driving a gas turbine (Figure 6a). A decisive factor is using the right technology to transfer the concentrated solar radiation through a glass window into the pressure vessel of the receiver (Figure 6b). Since the diameter of such heat-resistant quartz glass windows is limited by their fabrication process, a number of such modules are arranged in a matrix with conical mirrors (secondary

FIG. 6 | AN IMPROVED CENTRAL TOWER POWER PLANT

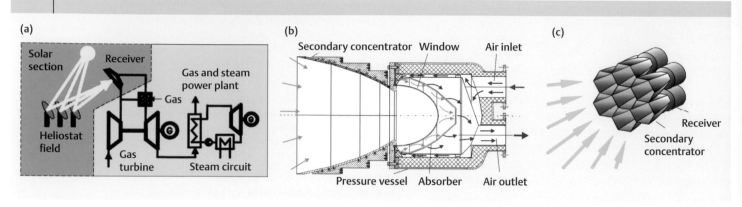

(a) Schematic of a solar combi power plant; (b) Layout of a high-temperature receiver module; (c) The conical mirrors of a number of modules can together make effective use of the concentrated radiation even with a large focal-spot area.

concentrators) in front of their entrance windows. These mirrors are shaped so that together, they form a large entrance aperture with practically no gaps (Figure 6c).

In an experiment at the Plataforma Solar, thus far three such modules have been combined and connected to a small 250 kW gas turbine. They produce temperatures up to 1030° C at a pressure of 15 bar. In early 2003, the turbine generated electric power for the first time. This represented an important milestone on the road to a large-scale technical application. It has led an industrial consortium to plan a demonstration plant in Spain with an output of 5 MW$_e$. The researchers expect a reduction of up to 20 % in power generating costs from this concept.

An additional important component which can contribute to cost reduction is the thermal energy storage reservoir. When a solar thermal power plant is operated on solar energy alone, the duty cycle of the power generating block which it drives is equivalent at a favorable site to an annual full-time operation of up to 2,500 hours. This duty cycle could be considerably increased if it were possible to store the energy from the solar field in a cost-effective manner. Then, the power plant could be equipped with a second collector field of the same size as the first, whose collected solar energy would flow into the storage reservoir. At times with little or no sunlight, the power generating block would then make use of this stored energy.

This increase in duty cycle would save the cost of investment for a second power generating block. The precondition is of course that the costs of the thermal energy storage reservoir are less than the additional cost of a larger power generating block. From present knowledge, this appears to be feasible. Cost-effective thermal energy storage concepts promise a further reduction in power-generating costs, which could again be up to 20 %.

Such a thermal energy storage reservoir would also have additional advantages. With it, power could be generated according to grid requirements, i.e. at peak demand periods. The price paid per kWh is then highest. It is also a plus on the technical side that the power plant would always op-

erate under optimal load conditions and could thereby minimize its heating-up and cooling-down losses.

The development of storage systems was long neglected in Europe: initially, the use of fossil fuels for bridging over periods with low sunlight was seen as the cheapest alternative – at least as a first step. However, it has the disadvantage that many subsidy arrangements do not permit hybrid operation (for example laws governing the subsidized feed-in of power to the grid).

A system currently in use for the parabolic-trough collectors with operating temperatures up to 400° C will permit intermediate storage of the heat energy in large tanks containing molten salts, which can be heated for intermediate storage by heat exchange with the thermal oil when the heat is not all required for steam generation. An alternative concept makes use of large blocks of high-temperature concrete as intermediate thermal storage reservoirs. With central-receiver systems, depending on the heat-transfer medium, there are two types of storage reservoirs. One type uses tanks containing molten salts; the other, useful when high-temperature air is the thermal transfer medium, passes the heat from the hot air to piles of small solid particles, which allow the air to pass between the particles and offer a large surface area, e.g. ceramic balls or quartz sand. The use of higher temperatures has a significant impact on the cost of storage systems, as they require less volume to store the same amount of heat energy. Heat losses from large storage reservoirs are very low due to the high volume-to-surface ratio, and overall they amount to less than 5 % of the annually stored thermal energy.

The Lowest CO$_2$ Emissions

Solar thermal power plants are an important intermediate link between today's energy supply based on fossil fuels and a future solar energy economy, since they incorporate important characteristics of both systems. They have the potential of supplying the world's electrical energy requirements several times over from solar fields, and by means of simple storage methods, they can potentially deliver power as needed, in contrast to other sustainable energy sources such as wind energy.

Solar thermal power plants are also favored by the fact that they can reduce CO_2 emissions in a particularly effective way. This becomes clear if one sums up the emissions which are due to the fabrication of the components, construction, operation and decommissioning of the plant via life-cycle analyses. Comparing various sustainable energy sources in this way, one finds the following balance for the specific CO_2 emissions per MWh of electrical energy generated: For solar thermal power plants, only 12 kg CO_2/MWh are emitted, while hydroelectric power plants emit 14 kg, wind-energy plants emit 17 kg, and photovoltaic power plants emit up to 110 kg CO_2/MWh [5,11].

Some of the photovoltaic modules are so unfavorable in this comparison because their manufacture is very energy-intensive and therefore causes a large amount of emissions. Advanced concepts require much less semiconductor material and therefore give significantly better emissions values. By comparison: Modern gas and steam turbine power plants emit 435 kg of CO_2/MWh, and coal-fired power plants as much as 900 kg CO_2 per MWh generated. These emissions are mainly due to the combustion of the fossil fuels.

For these reasons, different energy scenarios, for example that of the International Energy Agency (IEA) [6], predict that solar thermal power plants will be increasingly installed within the earth's Sun Belt – especially in the USA, Africa, India and the Middle East. According to this scenario, in some countries, solar thermal plants could contribute a significant fraction of the electric power consumed by the year 2050, up to 40 %. Furthermore, if a portion of this solar electrical energy is transmitted to neighboring industrial regions via high-voltage transmission lines, then by 2050, up to 10 % of the total power requirements could be provided. These estimates indicate that worldwide, by the year 2025 as much as 200 GW of electrical generating capacity from solar-thermal energy could be installed.

Summary

Solar thermal power plants collect sunlight and use its energy to drive thermal engines for electric power generation. Systems with steerable, silvered parabolic troughs which concentrate the solar radiation onto a central absorber tube, through which a heat-transfer medium flows, are already in commercial operation. In the central-receiver systems, a field of movable mirrors focusses the sunlight onto the top of a tower; a receiver there passes the heat energy to a thermal transfer medium. For small, decentral applications, dish-Stirling systems are suitable. These are steerable, paraboloid mirror dishes with a Stirling motor at their focal points. Thermal energy storage systems can be integrated and provide a cheap option for supplying flexible energy on demand as an added value for the power grid.

References

[1] *Sustainables Information 2003* – **2003** Edition, Publisher IEA, 201 pages (Jouve, Paris 2003).

[2] P. Heering, *Physik in unserer Zeit* **2003**, *34* (3), 143.

[3] G. Stadermann, Ed., FVS Themen 2002, *Solare Kraftwerke*, Forschungsverbund Sonnenenergie, Berlin **2002**.

[4] J. Solar Energy Eng., **2002**, *124* (5), 97; Special Edition.

[5] J. Nitsch *et al.*, Schlüsseltechnologie Regenerative Energien, Table 10.8, www.dlr.de/tt/Portaldata/41/Resources/dokumente/institut/system/publications/HGF-Text_TeilA.pdf.

[6] IEA Technology Roadmap: *Concentrating Solar Power*, www.iea.org/papers/2010/csp_roadmap.pdf.

[7] R. Pitz-Paal, in T.M. Letcher (Ed.), *Energy: Improved, Sustainable and Clean Options for our Planet*, Elsevier, **2008**.

[8] R. Pitz-Paal **2007**, in J. Blanco Galvez and S.Malato Rogriguez, Solar Energy Conversion and Photoenergy Systems, *Encyclopedia of Life Support Systems* (EOLSS), developed under the auspices of UNESCO, Eolss Publishers, Oxford, UK; www.eolss.net.

[9] A.T. Kearney and ESTELA, 2010, *Solar Thermal Electricity 2025*, ESTELA, June **2010**; bit.ly/OuYyQ3.

[10] *Concentrating Solar Power: Its potential contribution to a sustainable energy future*, ISBN 978-3-8047-2944-5, © German Academy of Sciences Leopoldina **2011**; www.easac.eu/fileadmin/Reports/Easac_CSP_Web-Final.pdf.

[11] J. Burkhardt, G. Heath, and C. Turchi, *Life Cycle Assessment of a Parabolic Trough Concentrating Solar Power Plant and the Impacts of Key Design Alternatives*, Environmental Science and Technology **2011**, *45(6)*, 2457; pubs.acs.org/doi/abs/10.1021/es1033266.

[12] DLR, **2006**. *TRANS/CSP: Trans-Mediterranean Interconnection for Concentrating Solar Power*, Final report for the study commissioned by the Federal Ministry for the Environment, Nature Conservation and Nuclear Safety, Germany; www.dlr.de/tt/trans-csp.

[13] S. Nage., M. Fürsch, C. Jägemann, and M. Bettzüge, *The economic value of storage in renewable power systems – the case of thermal energy storage in concentrating solar power plants*. EWI Working Paper no. 11/08, **2011**; www.ewi.uni-koeln.de/publikationen/working-paper.

[14] K. Ummel and D. Wheeler, *Desert power: the economics of solar thermal electricity for Europe, North Africa and the Middle East*, Center for Global Development, Working Paper no. 156, Dec. **2008**; www.cgdev.org/files/1417884_file_Desert_Power_FINAL_WEB.pdf.

About the Author

Robert Pitz-Paal received his doctorate in Mechanical Engineering from the University of Bochum. Since 1993, he has carried out research on solar energy in Cologne-Porz, and is currently co-director of the Institute for Solar Research and Professor at the RWTH (Technical University) in Aachen.

Address: *Prof. Dr.-Ing. Robert Pitz-Paal, Deutsches Zentrum für Luft- und Raumfahrt e.V., Institut für Solarforschung, Linder Höhe, 51147 Köln, Germany. e-mail: Robert.Pitz-Paal@dlr.de*

Photovoltaic Energy Conversion
Solar Cells – an Overview

BY ROLAND WENGENMAYR

The market share of photovoltaic power is growing rapidly. However, its contribution to the overall generation of electric power is still small. Photovoltaic systems are at present a mature technology and are long-lived, but they are still too expensive. This is however changing, thanks in part to new materials and production methods.

Photovoltaic energy conversion is convincingly elegant, as it transforms the energy of sunlight directly into electrical energy. Since the American inventor Charles Fritts fabricated the first selenium cells in 1883 (see Table 1), solar cells have developed into a technology which is mature and reliable from the users' point of view. At present, silicon technology dominates the production of solar modules, with around 90% market share. Alternative materials, in particular thin-film technology, are however gaining ground. Silicon is the workhorse material in the electronics industry, and its properties have therefore been thoroughly investigated, while its industrial process technology is well established and supplies of raw material are practically unlimited.

Only rather recently have the solar-cell producers emerged from their niche in the shadow of the all-powerful chip manufacturers. With an increasing economic leverage, these young companies can bring new technologies to the market, which are more suited for photo-voltaic applications. The current fabrication methods, in particular, still exhibit excessive losses of valuable semiconductor material. Wafers of monocrystalline silicon, the preferred material up to now, are the starting material for the production of solar modules. They are cut out of costly single-crystal blocks ("ingots"). With polycrystalline silicon, corresponding discs containing many small crystals are cut out of silicon blocks consisting of many small crystallites. These blocks are either cast from molten silicon in crucibles, or else the material in the crucible is melted by induction heating using strong electromagnetic fields.

The step of sawing out the wafers reduces much of the starting material to powder, both in the case of monocrystalline and of polycrystalline silicon. In the process, a considerable amount of energy is wasted, which was invested in growing the crystal or in melting the multicrystalline ingots. This worsens the energy balance and increases the production costs. Only through new technologies and materials, which are presently in various stages of development from research through pilot projects to commercial processes, can this fundamental problem of the conventional photovoltaic industry be eliminated.

In the production and application of photovoltaic modules, Germany, following China and Taiwan, belongs among the three leading countries. In spite of growth rates of around 30 % per year in Germany, the market, considered in absolute terms, is still small. In 2009, the total installed photovoltaic power in Germany was 9,785 megawatts of peak power (MW_p) [1]. These photovoltaic systems generated all together 6.6 terawatt hours of electrical energy in 2009 [1]. This corresponds to just under 1.2 % of the overall electric power consumption in Germany [2].

Within the solar industry, by the way, the power of solar modules is usually defined in terms of watts/peak (W_p), or peak power. This corresponds to normalized test conditions and is used to compare different modules. It however does not necessarily represent day-to-day operating conditions, and also does not indicate the power produced by the module when the sunlight is at its strongest.

Harvest factors are still too low

Especially when compared to wind and hydroelectric power, photovoltaic power conversion is still very much in the background, in spite of considerable government subsidies.

TAB. 1 | HISTORICAL MILESTONES

Year	Type of Solar Cell	Efficiency	Developer
1883	selenium (photocell)	nearly 1 %	Charles Fritts
1953/4	monocrystalline silicon	4.5–6 %	Bell Labs, USA
1957	monocrystalline silicon	8 %	Hoffmann Electronics, USA
1958	monocrystalline silicon	9 %	Hoffmann Electronics, USA
1959	monocrystalline silicon	10 %	Hoffmann Electronics, USA
1960	monocrystalline silicon	14 %	Hoffmann Electronics, USA
1976	amorphous silicon	1.1 %	RCA Laboratories, USA
1985	monocrystalline silicon	20 %	University of New South Wales, Australia
1994	gallium indium phosphide/ gallium arsenide, with concentrator	over 30 %	National Sustainable Energy Lab (NREL), USA
1996	photoelectrochemical, Grätzel cell	11.2 %	ETH Lausanne, Switzerland
2003	CIS, thin-film cell	19.2 %	NREL, USA
2004	polycrystalline silicon	20.3 %	Fraunhofer ISE, Freiburg, Germany
2009	gallium indium phosphide/ gallium gallium indium arsenide/ germanium; tandem cells	41.1 %	Fraunhofer ISE

Renewable Energy. Edited by R. Wengenmayr, Th. Bührke. Copyright © 2013 WILEY-VCH Verlag GmbH & Co. KGaA, Weinheim

Fig. 1 *The glass roof of the Stillwell Avenue station on the New York subway in Brooklyn contains the largest thin-film solar-cell installation in architectural settings, with a nominal power output of 200 kW$_p$* (photo: Schott).

Owing to the waste of material in the currently-used production processes, it is up to now also inferior to the other sustainable energy sources from the ecological point of view. This can be seen by taking a look at the "harvest factor": It represents the useful energy that a plant generates in the course of its lifetime in relation to the energy that was required for its construction and installation.

This mean harvest factor is in the range of 2 to 38 for photovoltaic plants, on the average around 10. With a life expectancy of 20 to 40 years, the plant will thus have been amortized in terms of energy after three years of operation; thereafter, it will generate more technically usable electrical energy than was required for its construction. Wind energy, in contrast, yields harvest factors between 10 and 50, while large hydroelectric plants, due to their long life cycles, have values up to as high as 250. In the case of coal-fired power plants, the harvest factor is about 90, and for nuclear power plants, it lies between 160 and 240; the energy cost of mining, transporting and consuming the fuel is included in these figures.

The energy and materials costs are still high, but are decreasing dramatically due in part to the increasing demand for photovoltaic systems. According to the EU Commission, the consumer cost of solar modules decreased by nearly half between 2007 and 2009. The fall in prices accelerated relative to past trends accordingly. In Germany, in addition, the cost of solar power, which all power consumers have to bear owing to the subsidies for photovoltaic plants through the Sustainable Energies Act (EEG), is undergoing massive decreases. According to a study released by the Fraunhofer Institute for Solar Energy Research (ISE) in Freiburg, Germany, the price of electric power from photovoltaic plants will fall below that from offshore wind-energy parks as early as 2013 (Figure 2) [3].

In the future, photovoltaic plants can offer enormous opportunities. Not only does the threatening climate collapse provide a weighty argument in favor of strong continuing support for research and development of photovoltaic devices; photovoltaic power generation is still a young technology as a large-scale energy source, even though it has been in use in space vehicles since the 1960's, and it still has strong potential for further development.

Solar cells can also offer some tangible plus points in terms of decentralized energy supplies. Windows or glass roofs coated using the new thin-film processes can combine electric power production in an elegant manner with the necessary light and heat management, especially for large buildings (Figure 1). This development is still in its infancy. It is also probable that in view of the increasing global energy demand, the price of electric power will continue to rise and this will make photovoltaic power generation economically competitive in the future.

FIG. 2 | **EVOLUTION OF POWER COSTS**

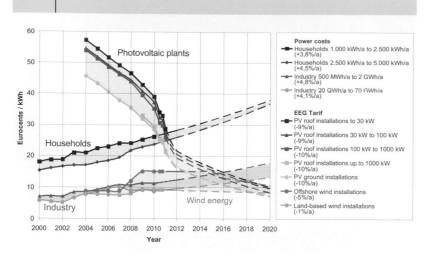

The evolution of power costs and tariffs from the Sustainable Energy Act in Germany shows that the costs of power from photovoltaic installations are sinking dramatically. As of 2013, it will cost the same as power from offshore wind-energy plants. From 2014 on, the bonus for power feed-in from photovoltaic installations will fall below the average industrial power cost (graphics: Fraunhofer ISE [3]).

Area-dependent and Area–independent Costs

The costs of photovoltaic plants are composed of a contribution which is proportional to the area covered by the solar modules, and a contribution which is independent of ths area. The latter includes for example the expensive power inverters, whose price has however decreased markedly in recent years through new technical developments. For plants which are to be connected to the power grid, the inverter converts the direct current from the solar modules, which are in a series circuit, into alternating current. If the power inverter is technically faulty or poorly matched to the solar modules, it can cause serious power losses and reduce the overall efficiency of the plant. In older installations, the power inverters were often a weak point. Modern units as a rule operate very reliably and, with proper installation, efficiently. When purchasing a power inverter, one should pay close attention to its conversion efficiency, since over the years, each percent lower efficiency accumulates in the form of noticeable losses. In 2009, researchers at the Fraunhofer ISE achieved an increase in the efficiency of power inverters to 99 % [4].

Among the area-dependent costs, besides the installation costs, the supporting frames and the cost of the land which may be a consideration, the price of the solar modules themselves is a significant factor. For this reason, researchers and developers are working steadily to further improve the efficiencies and reduce the fabrication costs of the solar modules. Progress is however difficult, although there is no lack of new concepts and materials combinations. The bottleneck on the way to the market is the industrial processing. Production plants are expensive and take some years to be amortized, and they have to yield modules of reliably high quality. Therefore, many producers of solar modules are

Fig. 3 *A FLATCON™ module from Freiburg, with concentrator cells (photo: Fraunhofer ISE, Freiburg).*

technically conservative. It requires considerable time before innovative processes reach commercial maturity.

As a result, monocrystalline and polycrystalline silicon will continue to dominate the market for some time to come, although silicon, as an indirect semiconductor, has notable disadvantages (see the infobox "Compact Fundamentals of Photovoltaics" on the next page). In 2002, cells made from polycrystalline wafers gained a larger market share than those made from the more expensive monocrystalline starting material for the first time. This represents the first material that was developed expressly for the photovoltaics industry. In the meantime, more wafers are now being produced for the photovoltaic industry than for electronics devices.

Since both these starting materials for solar modules suffer from a high materials loss during processing, the industry has for some time been searching for new methods which would avoid sawing the wafers out of a block with an inevitable waste of material. These new processes are collectively referred to as ribbon silicon or silicon foil. The raw material for the modules in this case is pulled directly from the silicon melt with the required thickness. Giso

Hahn describes these fascinating methods in the following chapter. Ribbon silicon has in the meantime captured its own small share of the market.

Modules fabricated from single-crystal silicon have thus far yielded the highest efficiencies. The so-called "black silicon" promises to yield still higher energy yields, since its surface captures an extremely large proportion of the incident light and hardly reflects any of it. This is achieved by a carpet of nano-needles which act like an artificial compound insect eye. This concept is however still at the stage of basic research. Commercial monocrystalline cells convert up to 22 % of the incident sunlight into electrical energy. Commercial poly-crystalline modules yield at best 17 % efficiency; for ribbon silicon or foil, the efficiencies are – depending on the fabrication method – 14 to 18 %.

Thin Films for Glass Facades

A different approach is used in processes in which the active photovoltaic layer is deposited as a very thin film onto the substrate material, usually glass. Such films can be made for example of cadmium telluride (CdTe) or copper indium disulfide (CIS). Both technologies offer the hope of considerable cost reductions and are treated in the following chapters. The efficiencies of commercially-produced cells lie in the range of 13 %, while in the laboratory, they attain up to 20 %. *Würth Solar* in the German city of Schwäbisch Hall started the first mass production of CIS solar modules worldwide in 2003, followed in 2005 by the Berlin firm *Soltecture* (previously called *Sulfurcell*; see the chapter after the next). The production capacity at *Würth* has meanwhile been increased to 350,000 solar modules per year [5]. This Swabian firm even offers them in the form of colorful façade claddings.

A related material is chalcopyrite (CIGS). It is a compound of copper, selenium and either indium, gallium or aluminum ($CuInSe_2$, $CuGaSe_2$, or $CuAlSe_2$). The German Center for Solar Energy and Hydrogen Research (ZSW) in Baden-Würtemberg currently holds the world record for the efficiencies of CIGS thin-film solar cells on glass, at 20.3 %. The efficiencies of commercially-produced cells are around 12 %. Swiss researchers recently presented a flexible CIGS cell on a plastic foil, which has an efficiency of nearly 19 %.

Silicon is also suitable for thin-film processing. The process consists of growing a thin layer of either microcrystalline or amorphous silicon onto a substrate material. Micro-crystalline means, in comparison to polycrystalline, that a large number of microscopically small silicon crystallites make up the layer. In an amorphous solid, in contrast, the atoms no longer exhibit a long-range spatial order over more than a few atomic radii. Amorphous silicon (a-Si) is enriched in hydrogen during its production process (a-Si:H). The hydrogen is able to partially "repair" the many defects in the disordered material and thus prevents the efficiency of the resulting solar cell from being reduced too drastically (see also the infobox "Compact Fundamentals of Photovoltaics").

Thin-film solar cells made from a-Si:H can be readily fabricated on relatively large areas of a glass substrate at a moderate cost. Commercial modules currently have typical efficiencies around 8 %, if they are required to be partially transparent as a glass coating. Up to 15 % efficiency is possible if they have a reflective coating behind the absorber layer. Thin-film modules of a-Si have the disadvantage that their efficiency at first drops off by 10 to 30 % of its initial value, before it stabilizes after a certain period of operation. On the other hand, they are much less sensitive to high operating temperatures than crystalline silicon solar cells. These lose about 0.4 % of their efficiency for every degree Celsius above their standard operating conditions (see the infobox "Determining the Efficiency" on p. 50); the efficiency loss of a-Si:H is in contrast only 0.1 %/degree. As a result, if the efficiency of a crystalline silicon solar cell at 25 °C is, for example, 18 %, then it is reduced at 50° C to

---

COMPACT FUNDAMENTALS OF PHOTOVOLTAICS

When a photovoltaic cell absorbs a light quantum (photon) from sunlight, that photon can raise the energy of an electron from the bound states ("valence band") within the solid. The electron leaves an empty "hole" in the crystal lattice. This hole can hop from atom to atom and thus can – like the excited electron – contribute to an electric current. However, this works only when the electron is excited at least into the region of the next higher allowed energy levels (the "conduction band"). In between, in solids there is an energetically forbidden zone. The photon must therefore be able to give sufficient energy to the electron to cross this zone.

The flow of current within the solar cell is reminiscent of the two-story Oakland Bay Bridge in San Francisco: on the upper story, the "electron traffic" flows in one direction, while on the lower story, the ""hole traffic" flows in the opposite direction. The cell obtains electrical energy from this bidirectional flow. However, this can function only when the two currents have predetermined flow directions to the corresponding electrodes, i.e. when the flows are strictly controlled along "one-way streets". This is guaranteed by two regions in the cell which act as valves to allow the flow only in the desired direction: the "p-doped region" lets holes pass through in one direction to the electrodes, while the "n-doped region" passes electrons in the opposite direction. The solar cell thus works like a rectifier or diode. In fact, it reverses the action of a light-emitting diode, in that it converts the energy of light quanta into electrical energy.

Important for the fundamentals of solar cells is the difference between direct and in-direct semiconductors. In direct semicon-ductors, such as gallium arsenide, the forbid-den energy zone is sufficiently narrow that photons from sunlight can lift the electrons directly across it and into the conduction band. Silicon, in contrast, is an indirect semicon-ductor. In this case, coupled vibrations of the atoms in the crystal (phonons) have to help in order to raise the electrons into the allowed energy region. Silicon is therefore a less favorable material for photovoltaic appli-cations. Only amorphous thin-film silicon is a direct semiconductor.

The excited electrons and holes must also have a long "lifetime", so that the electric current can flow in the cell with a minimum of perturbations. Defects in the crystal lattice allow some of the electrons and holes to "recombine" with each other; this is an un-wanted side effect which causes them to be lost for the production of electrical energy. This is also the reason why amorphous silicon has not long since attained a similar efficiency as crystalline silicon, although it is a direct semi-conductor.

Especially in new, less perfect materials such as ribbon silicon, a sophisti-cated "defect engineering" is necessary; this is de-scribed by Giso Hahn in the following chapter. A more detailed description of the physics of solar cells is given for example in [13].

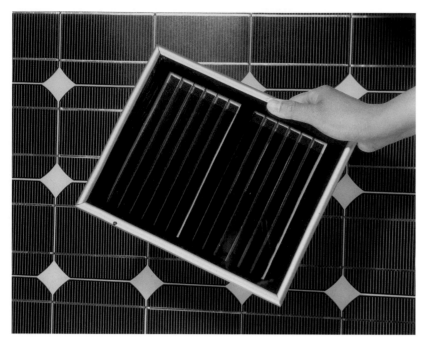

Fig. 4 *This photo-electro-chemical solar cell (dye solar cell) from the ETH Lausanne has 11 % efficiency and a high thermal stability* (photo: CH-Forschung).

15.5 % – these are quite realistic operating conditions. A cell made of a-Si:H would lose only one-fourth as much. Modules made from crystalline silicon must therefore be effectively cooled by an airflow. At very hot locations, a-Si:H can exhibit a superior energy yield.

Glass coated with a-Si has a lower efficiency, but for architectural applications, it has very attractive properties. The transparent modules can – along with power generation – also carry out the central function of active light and heat management. The currently largest thin-film solar installation based on a-Si is integrated into the roof of the Stillwell Avenue subway terminal in Brooklyn, New York. With a module area of about 7,000 m², this large New York subway station produces a maximum power of 210 kW$_p$ (Figure 1).

Highest Power Outputs

The attempt to increase the efficiencies of single-crystal silicon runs up against a theoretical limit of 28 % under standard conditions (cf. the following chapter). Another semiconductor material, gallium arsenide, permits a further increase, since it is a direct semiconductor (see the infobox "Compact Fundamentals of Photovoltaics"). Gallium arsenide is therefore often used in optoelectronics. It has, however, the disadvantage of being an expensive material. It also cannot be structured as readily as silicon using standard semiconductor techniques.

Even with gallium arsenide, efficiencies above 30 % can be obtained only if the solar cells are fabricated with a complex structure. An example is the tandem cell. Using a cell of this type, which has an especially high efficiency, researchers from the Fraunhofer Institute for Solar Energy Systems (ISE) in Freiburg recently attained a world-record efficiency of 41.1 %. Such cells combine different direct

semiconductor materials (see the infobox "Compact Fundamentals of Photovoltaics") in layers, each of which is optimized for one of three different spectral regions of the sunlight, and they therefore make use of the energy of the light much more effectively than conventional cells.

The layers consist of the materials gallium-indium phosphide and gallium-indium arscnidc on a substrate of gallium arsenide or germanium. The problem is that the crystal structures of these different materials do not match; during layer growth, mechanical stresses and crystal defects are produced, and these decrease the obtainable efficiency of the cell. The Fraunhofer researchers were able to solve this problem: Their cell "channels" the unavoidable stresses into a region where no light is absorbed, so that they do not disturb the operation of the cell. Moreover, the Freiburg cell achieved its record efficiency only with the aid of a lens above the cell, which focuses the sunlight onto the cell like a magnifying glass, and concentrates it there by a factor of 454. Only with the aid of this light concentration can these complex cells make optimal use of the light. This is the only way to make the specific-area price of the expensive cells competitive with conventional cells. The firm *Concentrix Solar*, a start-up from the Fraunhofer ISE in Freiburg, is for example already marketing concentrator cells for use in power-producing installations (Fig. 3).

Concentrator cells however require direct sunlight; therefore, they are usable only within the Sun Belt of the Earth. In Spain, Africa, Australia or the USA, they may one day even be cheaper than conventional cells, although they will require a tracking device to keep them pointed at the sun.

But even with conventional modules, a more adroit utilization of the light can permit increases in efficiency. Bifacial, or "two-faced" solar cells are for example transparent on their back sides at those places where they have no contacts. They can thus make use of light which falls onto the back side – e.g. via a mirror. Practically market-ready bifacial cells have already demonstrated an increase of their original efficiencies by up to one-fifth, and laboratory models have shown increases of more than fifty percent .

Organic, Polymer, Dye and Biological Solar Cells

Not only "hard" semiconductor materials can convert sunlight into electric power. Organic molecules in principle also have this property – mainly organic dyes. Some long-chain polymeric materials also behave as semiconductors and are in principle suitable for the fabrication of solar modules.

Such organic and polymeric solar cells are still at the stage of basic research. They are interesting as alternatives to conventional semiconductor materials because they have several attractive properties. For example, it is already clear that they would allow the production of cheap solar modules and would consume a comparatively small amount of energy during production. In contrast to the "hard" semiconductor materials such as silicon, they require no high temperatures for melting. In addition, their production can be made very environmentally friendly. In particular, however, they open up completely new possibilities, since they are light and flexible. For example, they could be integrated as colored solar cells into clothing, or used in architectural structures. Recently, researchers at the Massachusetts Institute of Technology in Cambridge, MA demonstrated that organic solar cells can even be evaporated onto a paper substrate – the first step towards solar wallpaper [6].

However, at present the organic and polymeric solar cells still have low efficiencies; in the laboratory, a special nanostructured material attained an efficiency of over 10 % for the first time [7]. Without such tricks, the best obtainable efficiencies lie in the range of 10-11 % [8]. The greatest problem for plastic solar cells is that they age rapidly under UV irradiation. Furthermore, their molecular lattices allow smaller molecules to pass through: Water and oxygen which penetrate the layers cause a rapid decrease in efficiency. Several smaller firms intend to introduce organic solar cells onto the commercial market, among them *Heliatek* in Dresden [8].

A further possibility for converting sunlight into electric power is offered by the photo-electrochemical cell, also known as the Grätzel cell. Michael Grätzel's group at the ETH Lausanne in Switzerland developed them in the early 1990's. They are also known under the term 'dye solar cells' or solar modules. Like a battery or an accumulator, these electrochemical cells contain an electrolyte, at present usually in the form of a gel, which is no longer liquid, as well as two electrodes. The electrolyte contains an organic dye whose molecules trap photons from sunlight and set electrons free as a result. In combination with the electrodes, the cell can then produce electric power.

The Lausanne group increased the efficiency of their cells in the course of the 1990's up to around 11 %. Currently-produced cells also exhibit a high degree of thermal stability, which is important for their practical application. In a long-term trial of 1,000 hours at 80° C, the Lausanne cells lost only 6 % of their original efficiency [9] (Figure 4).

The researchers at the Fraunhofer ISE are proceeding in a different direction. They have developed a special silkscreen printing process which allows them to make the area of dye modules especially large. The challenge here is to guarantee an effective long-term encapsulation of the electrolyte. Currently, they have demonstrated the development of a module 60 × 100 cm in area (Figure 5). Its efficiency could be increased to over 7 % [10].

The work of the Lausanne group shows that dye solar modules can be developed to provide serious competition for conventional solar cells in the area of large-scale power generation. A future development could be a variant of the Grätzel cell which employs a chlorophyll-like molecule to carry out artificial photosynthesis, mimicking the plant world.

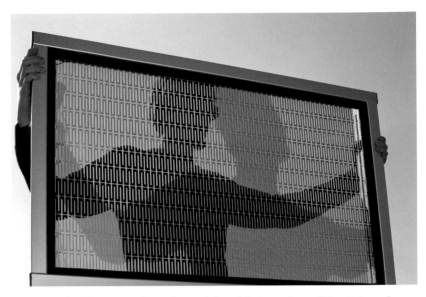

Fig. 5 *The first large-area dye solar module, with dimensions of 60 × 100 cm, from the Freiburg Institute for Solar Energy Systems, made by silk-screen printing on glass using a special technology* (photo: Fraunhofer ISE).

Physicists at the Technical University in Ilmenau, Germany, for example are carrying out research on such "biological solar cells". They can in particular have high efficiencies when the sunlight is weak, and could thus make more effective use of daylight under a variety of conditions. However, it will take some years of work before such cells are ready to be commercially produced and take over a share of the photovoltaic market. In the near term, organic, polymer and dye solar cells will remain exotic alternatives which still require considerable research and development before they are ready for large-scale application.

Suggestions for Planning a Solar Installation

In the planning of a photovoltaic installation, some practical questions have high priority. They include the annual insolation at the location of the installation, the optimal arrangement of the installation and its components, and questions of financing and availability of possible subsidies. Whoever wants to invest in a small installation for the roof must think in longer terms. Especially in the case of small installations, it can take more than 10 years for the installation to be amortized. This is strongly dependent on the region and the precise location of the installation. A decisive factor for an installation in Germany is also the rate of subsidies provided by the Sustainable Energy Act. According to its latest amendment, the subsidies were decreased in all cases of small rooftop installations by 9 % per year. In addition, there are supplements – or deductions – which depend on the currently-installed nominal power output of the installation [11]. They are intended to prevent cost explosions for consumers of electric power.

In Southern Germany and Switzerland (Mittelland), a well-planned and correctly installed setup with 1 kW peak power, corresponding to around 8 m² of module area, will provide an annual electrical energy yield of 900-1100 kWh. Further north, the yield will decrease. However, a study carried out in Freiburg, Germany, showed that in many cases, incorrect installation prevented obtaining the optimal yield [12]. Often, the modules are not correctly pointed to receive maximum sunlight, or are even in shadow some of the time, e.g. because of trees.

If the estimated yield provided by the producer of the installation is felt to be untrustworthy, an independent engineering opinion can be obtained. The TÜV (Technical Monitoring Agency) in Hessia for example offers such a service, with measurements at the planned location, on a cost basis.

A source of problems which is more difficult to pin down is technical errors in the fabrication of the modules. Some of the solar modules investigated produced less than their nominal power, or the power inverter was not optimized for the modules used or was even itself defective.

Whoever plans to operate such an installation should therefore pay attention to these possible problems. In the case of a newly-built house with a rooftop solar installation, the architect and the contractor for the solar installation should plan it together from the outset in order to obtain optimal results.

An important topic is also fire protection. Fire departments are increasingly demanding that in case of a fire, they must be able to shut down the photovoltaic installation rapidly in order to protect themselves. The best solution to this problem is the installation of an easily-recognizable "firemen's switch" near the entrance to the property.

Summary

The market share of photovoltaic power generation is growing rapidly. But in absolute terms, its contribution to providing electrical energy is still very limited. The advantage of a long lifetime for a photovoltaic installation is offset by its high initial investment costs. An increasing market share, new materials, and new technologies, e.g. thin-film solar cells, will however lower those costs in the future. It is important for optimal performance to plan and implement the installation correctly.

References

[1] Sustainable Energies in Numbers– national and international developments, Update: Dec. 2010. Bundesministerium für Umwelt, Naturschutz und Reaktorsicherheit, Berlin **2010**. http://bit.ly/KQmpFN.

[2] Annual Photovoltaics Status Report 2010, European Commission, bit.ly/dxPhWu.

[3] Skizze eines Energieentwicklungspfades basierend auf erneuerbaren Energien für Baden-Württemberg (Sketch of an energy development scenario based on sustainable energies in Baden-Württemberg). Fraunhofer ISE, Freiburg **2010**. bit.ly/ig5n3F.

[4] Press release No. 17/11 from the Fraunhofer ISE, May 26, **2011**. bit.ly/OW1Oaj.

[5] www.wuerth-solar.de/solar/de/solar/en/wuerth_solar_2012.

[6] M.C. Barr *et al.*, Adv. Mat. **2011**, online 8th July, DOI: 10.1002/adma.201101263.

[7] X. Dang *et al.*, Nature Nanotech. **2011**, 6, 377.

[8] www.heliatek.com/?lang=en.

[9] CH-Forschung 2002, www.ch-forschung.ch.

[10] Press release No. 5/11 from the Fraunhofer ISE, March 10, **2011**, bit.ly/P4dw2S.

[11] www.bmu.de/english/aktuell/4152.php, en.wikipedia.org/wiki/German_Renewable_Energy_Act.

[12] Press releases of the Fraunhofer ISE and the *Badenova AG* on 1. February 2006, www.ise.fhg.de/de/presse-und-medien, Presseinfos **2006**.

[13] P. Würfel, U. Würfel, Physics of Solar Cells, 2nd Ed. Wiley-VCH, Berlin **2009**.

About the Author

Roland Wengenmayr is the editor of the German physics magazine "Physik in unserer Zeit" and is a science journalist.

Address:
Roland Wengenmayr, Physik in unserer Zeit, Konrad-Glatt-Str. 17, 65929 Frankfurt, Germany Roland@roland-wengenmayr.de

INTERNET

Solar-cell research worldwide:
Many research groups are active in the field of photovoltaics in universities and research institutions around the world, so this is only a very brief selection
www.nrel.gov/solar
sfc.mit.edu
photonics.stanford.edu/research/working-groups/solar-cell
cleantechnica.com (US portal)
sun.anu.edu.au
www.aist.go.jp/aist_e/aist_laboratories/4environment/index.html

Solar-cell research in Germany (selection)
www.fvee.de
www.ise.fraunhofer.de/en?set_language=en
www.fz-juelich.de/portal/EN/Research/EnergyEnvironment/_node.html
www.helmholtz-berlin.de/forschung/enma/index_en.html
www.isfh.de/?dm=1&_l=1
www.iset.uni-kassel.de
www.uni-konstanz.de/photovoltaics

International studies on the long-term behavior of photovoltaic installations
www.iea-pvps-task2.org

Solar Cells from Ribbon Silicon

BY GISO HAHN

The solar-cell market is booming. But photovoltaic cells are still too expensive to be able to compete effectively with conventional power generation. A notable cost reduction can be achieved if ribbon silicon is used instead of the usual silicon wafers, which are sawed from massive blocks of silicon.

The term *photovoltaic* energy, derived from the Greek, can be translated descriptively as "electrical energy from light". In the year 2003, the solar cell, which is at the heart of every photovoltaic module, celebrated its 50th birthday. There are many reasons for the increasing success of solar power. Although in the beginning, satellites were the main users of solar modules as an independent source of electric power, photovoltaic energy soon came down to earth. In terrestrial applications, a number of advantages play a role: For one thing, photovoltaic (PV) power permits a sustainable energy supply within closed systems, and thus provides freedom from the limited supplies of fossil fuels and their negative effects on the environment. A second advantage is the possibility of a decentralized energy supply for isolated sites with no connection to the power grid, e.g. for mountain cabins, traffic signs or settlements far from power lines. Not least, the modular character of photo-

voltaic systems is an important reason for the optimistic prognosis regarding the future development of this form of energy conversion.

However, the relatively high energy production costs of power generated by photo-voltaic modules has so far put the brakes on the growth of this technology for earthbound applications. While conventional base-load power plants can generate power today at prices between 1.5 and 3 €-cent/kWh, and peak-load gas turbine plants cost between 8 and 11 €-cent/kWh, the generating costs for photovoltaic power are currently much higher. In Germany, the "100,000 Roofs Programme" and the "Renewable Energies Act" (EEG) have given photovoltaic power generation a strong boost since 1999: The EEG subsidizes the feed-in of PV power into the power grid at up to 20 €-cent/kWh for up to twenty years with installations put in service by the summer of 2011. The subsidies are lower for installations put in service in later years. If research and industry succeed in making effective use of the existing potential for savings in the production of photovoltaic solar modules, they may – thanks to these subsidies and the positive effects of mass production – obtain a further substantial cost reduction that would make photovoltaics economically competitive with conventional power generation even without subventions. While the price of photovoltaic-generated power is essentially determined by the level of subsidies, the true costs (not including the profits of the manufacturing companies) are lower.

The State of the Art

The annual growth rates of installed electric power generated by photovoltaic systems in the past few years were around 30–60 %. The leading nation in sales of photovoltaic modules is currently China, ahead of Taiwan, Japan and Germany (see Fig. 1). In the year 2011, new solar cells with a power output of over 37,000 MW were fabricated worldwide. By comparison, a nuclear power plant has an output power of 1,000–1,500 MW. Since photovoltaic installations generate power only when the sun is shining, a realistic comparison of the energy produced by the two technologies within a given year must take a conversion factor into account. If growth rates continue at the same level, it will thus take some time yet until solar power contributes a substantial portion –5 to 10 % – of the total worldwide installed electric power generating capacity. In Germany, today already about 4 % of the overall electric power requirements are provided by photovoltaics. To accelerate this growth, it

A bank of furnaces from which silicon ribbons are being pulled from the melt in the form of hollow, octagonal columns up to 5 meters high, using the EFG process (photo: Schott Solar GmbH).

FIG. 1 | THE MARKET FOR PHOTOVOLTAIC MODULES

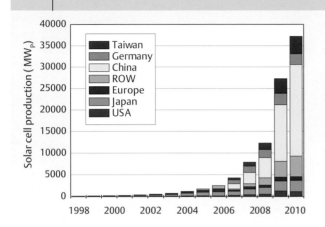

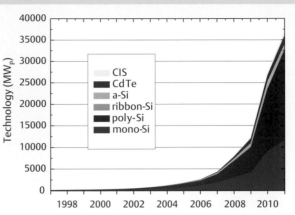

The production of photovoltaic modules by regions (left) and in terms of the various technologies which are currently economically relevant (right) [1]. A shift from single-crystal Si towards polycrystalline Si can be clearly seen.

is necessary to press on with research into various concepts that promise to reduce the cost of solar power as far as possible. The benchmark is the specific generation cost of one watt of PV power under standard conditions of solar radiation (see infobox "Determining the Efficiency", p. 50); this so called watt-peak cost (W_p cost) has to be reduced.

The first solar cells were fabricated from single crystals of semiconductor materials of the same quality as those used in the production of integrated-circuit electronic devices. An important factor in determining the suitability of a semiconductor material for photovoltaic applications is the size of its band-gap energy (see infobox "Solar Cells from Crystalline Silicon", p. 48). Gallium arscnidc (GaAs) is very suitable, since its band-gap energy of 1.42 eV is perfectly adapted to the solar spectrum. The band gap of silicon (Si), however, is only 1.12 eV, and is therefore somewhat too small. On the other hand, silicon can be produced with the required purity at a much lower cost. Solar power from silicon cells profits here from the years of experience with Si gained in the microelectronics industry. Furthermore, the waste products from the semiconductor industry in the past represented an important source of raw material, since the purity requirements for the fabrication of solar cells are somewhat less stringent than those for microelectronics. For these reasons, GaAs has been used primarily in space applications, where a high efficiency is decisive and production costs are less important. In terrestrial applications, in contrast, reduction of the W_p cost has been the main goal from the very beginning. For some time now, however, the demand for Si material by the photovoltaics industry has exceeded that of the microelectronics industry, so that in the meantime, the former has built up its own source of silicon supply.

For these reasons, single-crystal silicon (mono-Si) dominated photovoltaic appli-cations during the 1970s and 1980s. Poly- or multicrystalline silicon (poly-Si) later provided further cost savings. These have to be balanced against the lower conversion efficiencies of cells made from poly-Si, due to crystal defects such as grain boundaries be-

tween the individual crystallites, dislocations in the crystal lattice, and its higher impurity concentration as compared to mono-Si. The cheaper starting material however more than outweighs the loss in efficiency. Under suitable process conditions, the W_p costs of poly-Si can be markedly lower than those of mono-Si. As a result, poly-Si has in recent years displaced mono-Si from its predominant position and is now the leader, providing 57 % of annual installed power [1].

A common feature of both materials is that the flat discs ("wafers") needed for the production of solar cells are as a rule cut out of massive Si blocks ("ingots"). The ingot is first cast; for this process, the highly pure (and therefore expensive) silicon starting material has to be melted (melting point 1414°C) and then allowed to solidify under a well-defined temperature gradient. For an ingot weighing several hundred kg, this

INTERNET

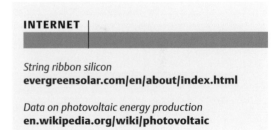

String ribbon silicon
evergreensolar.com/en/about/index.html

Data on photovoltaic energy production
en.wikipedia.org/wiki/photovoltaic

takes two to three days. Cutting the individual wafers out of the ingot is accomplished with wire saws, which use wires several kilometers in length and around 120 µm in diameter. The square wafers in final form are 180 to 200 µm thick and typically 156 mm on a side. Between 50 and 60 % of the starting material is lost in the course of fabricating wafers from the ingots; the major portion of this is literally pulverized in the process of sawing out the wafers! This increases the fraction of the production cost of a solar module due to wafer costs by 33 % [2] (see Fig. 2).

Thin-Film Solar Cells

For the reasons described above, there has for some time been a search for alternatives to crystalline silicon (c-Si), which is utilized in such a wasteful manner. One alternative is amorphous silicon (a-Si); in contrast to c-Si, it is a direct semiconductor. Direct semi-conductors absorb light much more effectively than indirect semiconductors; therefore, the active photovoltaic layer can be made much thinner. In

FIG. 2 | MODULE PRODUCTION COSTS

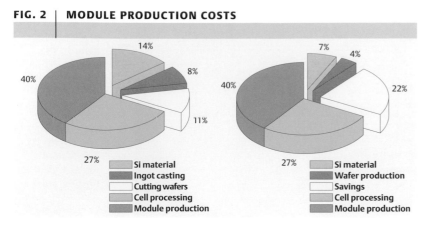

The cost distribution for solar modules. Left: poly-Si wafers sawed from ingots [2]. Right: Wafers of ribbon silicon with the same efficiency.

thin-film solar cells, this layer is only a few microns thick and contains only about one percent as much material as in the active layer of c-Si wafers. However, solar cells made from a-Si have the disadvantage that their conversion efficiencies degrade during their first thousand hours of operation. Their efficiency thereafter remains stable, but it is significantly lower than that of c-Si solar cells. This has a negative effect on the W_p costs, since they include other factors besides the production costs of the solar cells; these are area dependent and scale with the efficiency – for example the module costs, cost of the supporting frames, as well as direct costs for obtaining the site where the modules are to be set up. For these reasons, a-Si is applied successfully mainly for devices with low power-output requirements, e.g. for pocket calculators or watches, where the low production cost is important.

Other materials which are currently the subject of intensive research as absorbers for thin-film solar cells, and are already in production, include cadmium telluride (CdTe) and copper indium diselenide (CIS; see the following chapters). Although both materials have exhibited high efficiencies in laboratory experiments, the module efficiencies in industrial production are still notably lower than those of c-Si. Many people also consider the use of toxic materi-

als such as cadmium to be problematic, even though it is safely encapsulated in the solar modules. CdTe solar modules can currently be produced at a lower cost than modules using crystalline silicon, and have therefore captured a significant market share in spite of their lower efficiency.

Although intensive research has been carried out in the area of thin-film solar cells, the market share of photovoltaic modules based on c-Si has not changed significantly in the past few years (Figure 1). A further reason for this is the tendency of industrial producers to bank on an established technology in order to minimize risks. It is much simpler and more cost-effective for established producers to enlarge their existing production capacity than to introduce a completely new process technology. It is thus clear that crystalline silicon will continue to make by far the largest contribution to photovoltaic power generation throughout the coming decade.

Ribbon Silicon

A major portion of the costs of the current c-Si photovoltaic technology are due to the wasteful usage of expensive high-purity silicon. One possibility for reducing wafer costs consists of simply using thinner and therefore more readily breakable wafers. In principle, it is already possible to reduce the wafer thickness to 150 μm and thus to obtain more wafers from a cast ingot and reduce the cost per wafer, if the efficiency and the production yield (breakage) remain constant. This is however currently not the case. Furthermore, more 'sawdust' would be produced, so that the percentage of wasted silicon would increase.

Another, more elegant method is offered by the use of ribbon silicon. Silicon ribbons are crystalline wafers which are pulled directly from the melt with the required thickness of about 200 μm. Ribbon silicon wafers have the great advantage as compared to wafers cut from ingots that almost all of the material in the silicon melt can be used to produce the crystalline wafers. The lack of waste 'sawdust' gives a significant reduction in the fraction of wafer costs within the overall cost of the modules. In addition, a second cost factor in the conventional process can be eliminated completely, namely the crystallization of the silicon ingots. With the condition that the efficiency of the cells made from ribbon-silicon wafers be just as high as that of cells made from wafers sawed out of ingots, up to 22 % of the module costs can be saved by using ribbon-silicon wafers (see Fig. 2). This would be the case for so-called RGS silicon, which we will describe below.

Another important factor is the superior usage of silicon. In the past, the photovoltaic industry was able to fill its needs by using the excess production of silicon for micro-electronics. The strong growth of photovoltaics in recent times has, however, had the result that solar-power applications now use notably more silicon than the integrated-circuit producers. Conserving silicon material is thus an additional advantage in the use of silicon ribbon, if one is

FIG. 3 | TYPES OF RIBBON SILICON

The meniscus at the solid-liquid boundary is used to classify the preparation method for silicon ribbon [3]. M_1: pulling direction vertical, capillary forces push liquid Si up through a mold, after which it solidifies. M_2: pulling direction vertical, the ribbon is pulled from a broad base at the surface of the Si melt. M_3: pulling direction horizontal, the phase boundary extends over a large area.

concerned about maintaining the high growth rates of the photovoltaic industry.

Over the years, many different production technologies for ribbon silicon have been tried out. These technologies can be reasonably classified in terms of the meniscus formed by the Si melt at the phase boundary between the liquid and the solid phase (see Fig. 3). In the case of meniscus type M_1, the base area of the meniscus is defined by a mold into which the liquid Si climbs up above the surrounding surface due to capillary action. The wafer is pulled upwards out of the melt, at a pulling rate of around 1 to 2 cm/min; this is the main factor determining the thickness of the wafer. The heat of crystallization is carried off mainly by radiation, while convection hardly plays a role. Therefore, the rate of crystallization is relatively low and this, in turn, limits the pulling rates. During the pulling, the temperature gradient at the liquid-solid boundary must be controlled to within 1° C, i.e. with a very high precision from the technological point of view. This technique was commercialized as early as 1994 under the name 'Edge-defined Film-fed Growth' (EFG) and was being further developed up to 2009 by the *Wacker Schott Solar GmbH* company in Alzenau (Franconia) [4]. The silicon is pulled from the melt in the form of 5 m long tubes. A graphite mold gives it an octagonal shape with very thin walls. This closed shape avoids free edges which would have to be stabilized. A laser then cuts the 12.5 cm wide faces of the octagon into square or rectangular wafers.

The meniscus shape of type M_2 has a broader base than type M_1. It results when the wafer is pulled vertically upwards directly out of the melt. Owing to the longer meniscus, this type of silicon ribbon preparation can tolerate a greater fluctuation in the temperature gradient at the liquid-solid phase boundary, of around 10°C, which permits the use of more compact and less expensive processing equipment. An example of this type is the string-ribbon silicon developed by *Evergreen Solar Inc.* and commercialized since 2001. In this process, two fibers ('strings') made of a material which is kept secret by the producer are passed through the Si melt and pulled parallel and vertically upwards from the liquid surface (see infobox 'Internet' on p. 45). This process makes use of the fact that silicon has a high surface tension, even greater than that of mercury. Thus a silicon film stretches between the two strings,

which are about 8 cm apart, like a soap-bubble film, and it solidifies to give a ribbon. A laser then cuts this ribbon into wafers of the desired size. In this likewise vertical pulling method, the pulling velocity is limited for the same physical reasons as in the EFG process to 1–2 cm/min.

Considerably higher pulling velocities are possible if the wafer is pulled horizontally instead of vertically from the melt. This is done using the extended meniscus shape of type M_3. For horizontal pulling, a substrate can be used on which the silicon solidifies. This is e.g. the case in the Ribbon Growth on Substrate (RGS) silicon process. This method was originated by *Bayer AG*, and is still in the developmental stage. A belt carrying substrate plates moves under the crucible containing the silicon melt. The silicon is deposited onto the plates. As soon as the wafers have crystallized, they detach themselves from the substrate due to the difference in thermal expansion coefficients, so that the substrate plates are again freed up for the next pass. This process is distinguished by rapid heat dissipation through the substrate plates. Furthermore, the direction of crystallization is decoupled from the horizontal pulling direction; it runs up vertically from the cooler face of the substrate to the upper surface of the wafers. Both these effects permit high pulling velocities up to 10 cm/s, i.e. a 30- to 60-fold higher throughput than in the vertical methods. This enormously reduces the cost of wafer production.

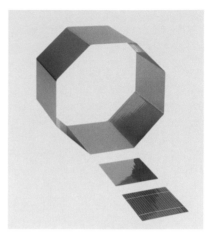

The octagonal tubes obtained from the EFG process are first cut into sections. Their faces are then cut into square wafers 15.6 cm wide (graphics: Schott Solar GmbH).

Crystal Defects and Defect Engineering

All of the types of ribbon-silicon wafers discussed here solidify in the form of polycrystalline silicon. Their particular process conditions lead to different types of lattice defects. We mention some typical kinds of defects using the examples of the three materials listed above: EFG silicon, string-ribbon silicon, and RGS silicon.

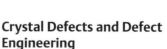

In the EFG process, the silicon ribbon is pulled out of the silicon melt in the form of hollow, octagonal columns up to over 5 meters high, whose walls are only about 300 μm thick (photo: Schott Solar GmbH).

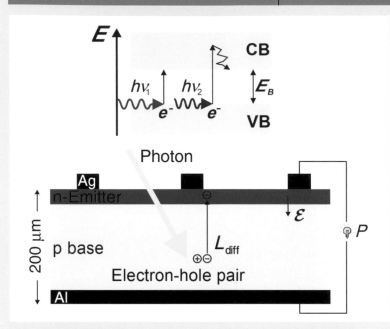

Above: Energy-band diagram. Below: The structure of a silicon solar cell, schematic; Ag denotes the silver front electrode, Al the aluminum rear electrode on the back side of the wafer.

A solar cell makes use of the internal photoelectric effect, in which incident photons remove electrons from their atomic bound states. In a semiconducting material, the photon must have at least the energy $h\nu_1$ which is sufficient to excite an electron out of the valence band (VB) over the band gap (of energy E_B) into the conduction band CB (upper part of the figure). In this process, an electron-hole pair is formed.

Photons with an energy $E = h\nu_2 > E_B$ excite electrons into states in the conduction band above the lower band edge. These then give up energy through collision processes until they have dropped back down to the band edge. The excess energy $E - E_B$ is thus lost in the form of heat. The required minimum energy E_B and the loss of excess energy reduce the maximum attainable efficiency of a solar cell made from crystalline silicon to about 43 %. Further loss mechanisms decrease the theoretically achievable efficiency down to 28 %.

The p-n junction in the solar cell produces an electric field (the 'built-in field', ε in the lower part of the figure). Mobile charge carriers are accelerated by this field in different directions depending on the sign of their charge. This gives rise to charge separation, so that a voltage appears between the emitter and the base electrodes. If the external circuit is completed through a power-consuming device, the electric power P can be extracted from the solar cell.

In the currently most common type of solar cells fabricated from c-Si, the absorber material must be sufficiently thick so that the photons are absorbed as completely as possible by the indirect silicon semiconductor. The absorption is stronger for light of short wavelengths (blue) than for light of long wavelengths (red). In order to use the long-wavelength part of the solar spectrum, and also for reasons of stability, such conventional solar cells are normally made about 180-200 µm thick. The charge carriers have to diffuse from the site where they are generated to the p-n junction in order to contribute to the current output of the cell. The diffusion length L_{diff} is the distance which the charge carriers can travel before they recombine. L_{diff} is related to the diffusion constant D and the lifetime τ of the charge carriers by the formula

$$L_{diff} = \sqrt{D\tau}.$$

In a high-quality solar cell, the charge carriers should have long lifetimes and thus a long diffusion length.

EFG and string-ribbon wafers contain long crystallites of a few cm² in area, oriented along the pulling direction, after their production; these are subdivided to some extent through twinning grain boundaries. The dislocation density in the different crystallites varies widely, leading to a very inhomogeneous material quality in the wafers. The main contaminant is carbon, with a concentration up to more than 10^{18} per cm³; it is thus considerably higher than in wafers which are sawed from Si ingots. Furthermore, metallic impurities may also reduce the wafer quality, however at much lower concentrations.

In the RGS wafers, in contrast, the much higher pulling rates lead to crystallite sizes of less than a millimeter. The dislocation densities, defined as the length of dislocations per unit volume, range up to 10^7 per cm². By comparison, the wafers used in the microelectronics industry are dislocation-free! The main portion of the impurities consists again of carbon, and at somewhat lower concentration, of oxygen.

Impurities and other crystal defects such as dislocations and grain boundaries have a decisive drawback: Because they disturb the translational symmetry of the perfect crystal, they reduce the lifetime of the charge carriers (Figure 4). The defects can form allowed energy levels within the band gap. These 'stepladders' permit an electron which has been excited into the conduction band by a photon to fall back down into the valence band and recombine. It is then lost for further charge transport. The goal is therefore to remove these defect states as far as possible during the fabrication of the solar cells. This increases the charge-carrier lifetimes and improves the quality of the finished solar cell. A longer charge-carrier lifetime for example increases the amount of current which the cell can deliver. More charge carriers reach the p-n junction and can be separated there by its electric field before they can recombine.

One strategy consists of removing defects, for example metallic impurity atoms, during the processing of the solar cells. This 'gettering' makes use of the fact that metals have lower solubilities in silicon than for example in aluminum. Aluminum comes into play in any case, as the rear contact of crystalline silicon solar cells: In their industrial production, the back side of the wafers is usually covered with an aluminum-containing paste. This is then burned into the wafer at a temperature of 800-900° C, forming the rear electrode. At these temperatures, most metal atoms are mobile in silicon; their higher solubilities in aluminum then cause an automatic purification of the silicon wafer, driven by the concentration gradient for diffusion. The metal atoms cause no problems in the aluminum layer, while the lifetime of the electrons in the purified, active silicon layer is lengthened. It is particularly attractive for the producers that this gettering does not require an additional process step, since electrode formation is already part of the production process of solar cells.

Of even greater importance for all types of solar cells made from poly-Si is the use of hydrogen. Atomic hydrogen can bind to defects such as dangling bonds in the crystalline Si lattice, which would otherwise give rise to impurity states within the band gap. The binding of hydrogen changes the bonds and bond angles and thus influences the energetic positions of the impurity states. This makes it possible to shift many of the impurity states within the band gap or to move them entirely out of the gap. The art of influencing lattice defects in a favorable way is known as defect engineering.

In mass production, a particularly elegant method of applying atomic hydrogen to the wafers has become prevalent in recent years. It makes use of the fact that the front surface of the wafers is covered with an antireflection coating in order to permit the maximum number of those photons arriving at the cell to enter it and produce charge separation. This is the reason for the characteristic deep blue color of the crystalline silicon solar cells. This function can be performed by a silicon nitride film, which is deposited from the gas phase using the PECVD method (Plasma-Enhanced Chemical Vapor Deposition). This film traps up to 20 at % of hydrogen during deposition. During forming of the back electrode, the high process temperature causes the hydrogen to diffuse into the silicon substrate, where it can bind to defect structures and shift the defect levels out of the band gap or at least into a more favorable position near the band-gap edge. When the recombination rate due to the defect levels is reduced in this manner, one refers to a *passivation* of the defects.

Gettering of impurities and passivation of crystal defects produce a great improvement in the quality of starting material, which contains a high defect concentration. This is especially true for ribbon-silicon material, whose quality can vary widely within small regions between neighboring crystallites. These two techniques make it possible to fabricate solar cells with good efficiencies from silicon ribbon (Figure 4). These process steps as a rule do not complicate the production process, since they are already a part of the processing sequence. However, they must be individually optimized for the specific material being used.

In Figure 5, the distribution of current within an RGS and an EFG solar cell is illustrated. While in the EFG material, the strip-like structures reflect the long crystallites which grow in the pulling direction, in the RGS cell in comparison one can recognize the lower current density and the smaller crystallite size.

Strategies for Cost Reduction

There are two strategies for reducing the cost of power production in photovoltaic devices: Increasing the efficiency of the modules, and lowering the production costs of the starting materials. In the past, research into solar cells was concentrated on increasing their efficiencies. In the meantime, it has become clear that it can make more sense economically to produce solar cells from cheaper material of lower quality, whose efficiencies in the end are only marginally lower than those of modules made from more ex-

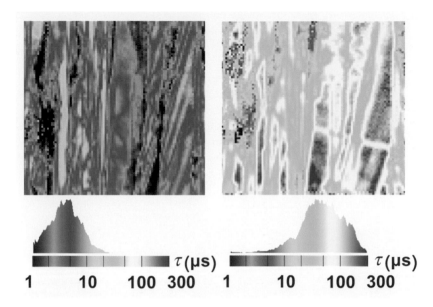

Fig. 4 *The distribution of charge-carrier lifetimes in a 5 × 5 cm wafer made of vertically-pulled ribbon silicon. Left: In its as-produced state just after preparation. Right: following gettering and hydrogen passivation. A lifetime of 10 µs corresponds to a diffusion length of about 170 µm.*

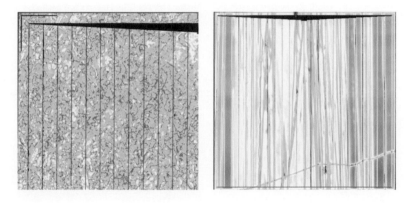

Fig. 5 *The electrical current distribution in solar cells. a) In RGS silicon, with small crystallites (the section shown corresponds to about 1 square cm). b) The current distribution in EFG silicon (four cm²); the current density increases from blue to green to red.*

pensive wafers. In such cost-benefit calculations, naturally other considerations based on area costs also play an important role, whereby the type of mounting and other factors are taken into account.

A promising candidate for a further notable W_p cost reduction is RGS silicon, because the high throughput of a single production installation and the good use of starting material can lead to the lowest unit costs for poly-Si wafers. The small crystallites and the high concentrations of carbon and oxygen are at present limiting factors to the efficiency of solar cells based on RGS silicon wafers. In particular, the charge-carrier lifetimes are still –even after gettering and hydrogen passivation steps – markedly shorter than in other types of silicon ribbons. Current research is therfore concentrating on ways to reduce these impurity concentrations. The initial efforts have already led to an increase in the charge-carrier lifetimes, which gives a corresponding improvement of the efficiencies of solar cells based on this material.

Research has discovered a very elegant method of collecting the major portion of the charge carriers in a material like RGS silicon, with its limited charge-carrier lifetime, before they are lost for current output due to premature recombination [5]. In the p-type base material, oxygen and carbon agglomerates can collect along extended crystal de-

fects such as dislocations. If these agglomerates are sufficiently densely packed along the defects, they form a sort of sheathing around the defect lines. Fixed positive charges along this sheath can then cause a local inversion of the charge-carrier type along the boundary layer, by repelling the majority charge carriers (holes in p-type silicon) and attracting the minority charge carriers (electrons). Thus, an n-type conducting channel is formed along the line of dislocations within the p-type silicon.

Thus, the RGS silicon wafer converts the disadvantage of poor crystal perfection into an advantage: The wafer is interlaced by a branched network of dislocations which reaches up to its surface and makes contact there with the n-type emitter layer. The short-lived electrons then no longer have to diffuse to the p-n junction near the surface after being generated by absorption of a photon deep within the cell; it suffices for them to reach the nearest n-type channel, through which they are quickly carried to the emitter layer at the device's surface, and can contribute there to the external current. Since the average spacing of the dislocation lines in RGS silicon is only a few micrometers, high currents can be obtained by this mechanism in spite of the short diffusion lengths (Figure 6). However, the extended space-charge region along the n-type network also causes higher recombination currents in the solar cell (diode saturation currents), so that the improved charge-carrier collection does not automatically lead to a correspondingly great improvement in overall cell efficiency.

FIG. 6 | CURRENT COLLECTION

A section through an RGS-silicon solar cell containing a network of n-type conducting dislocations. The network passes through the whole thickness of the wafer and is in contact with the n-conducting electrode at its surface. A bright contrast means a high current-collecting capacity (here made visible by the EBIC technique, Electron Beam Induced Current).

50 μm

Efficiencies

The efficiency, along with the production cost, is the decisive parameter of a solar cell. It determines the electrical power which the cell can produce. The theoretically attainable efficiency of solar cells based on crystalline silicon under standard conditions of solar irradiation, an AM 1.5 spectrum at 1000 W/m², is around 28 % (see the infobox "Determining the Efficiency"). The best efficiency which has been obtained in the laboratory for single-crystal silicon is 25 %, near this theoretical limit. However, the laboratory solar cells fabricated for this measurement are much too expensive for large-scale industrial production. The efficiencies of large-area solar cells which can be mass produced at low cost are currently much lower. For mono-Si, industrial efficiencies in the range of up to 19 % can be obtained (with more elaborately-structured cells, even above 22 %); for poly-Si, they range up to about 17.5 %.

With solar cells made of ribbon silicon, marked increases in efficiency have been demonstrated in recent years. For laboratory production, the highest efficiencies of string-ribbon and EFG cells are in the range of 18 %, and for RGS cells, 14–15 % has been demonstrated [6,7]. The highest values obtained using industrial-scale processing lie about 1–2 % lower (see Fig. 7). This dynamic evolution is driven in particular by the increasing knowledge of the materials properties and the development of processing techniques for solar cells which are adapted to them. New process sequences

DETERMINING THE EFFICIENCY

The efficiency of a solar cell is found from its current-voltage characteristic curve under illumination. This curve has the same shape as the characteristic of a diode, but is shifted along the current axis by the value of the short-circuit current I_{SC}. In order to take into account the area dependence, usually the current density j is quoted instead of the current I. The open-circuit voltage V_{OC} is the voltage which acts between the electrodes of the cell when no current is flowing.

The point of maximum power density P_{max} is reached when the product $j \cdot V$ is maximal. The yellow-green rectangle A_1 (Figure) with sides of length V_{OC} and j_{SC}, is thus larger than the red rectangle A_2 which lies above it and is defined by the point P_{max}. The ratio A_2/A_1 is called the filling factor FF. The efficiency η of a solar cell is defined as the ratio of P_{max} to the power density of the light impinging on the cell under irradiation, P_{in}:

$$\eta = \frac{P_{max}}{P_{in}} = \frac{V_{OC} \cdot j_{SC} \cdot FF}{P_{in}}.$$

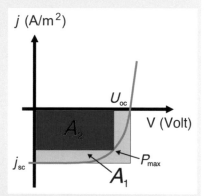

The current-voltage characteristic of a silicon solar cell.

For terrestrial applications, the efficiency is measured under standard conditions. In the AM 1.5 standard, the incident light spectrum has a power density P_{in} of 1000 W/m² and the cell is kept at a temperature of 25°C. This corresponds roughly to the average solar radiation at medium latitudes with a clear sky and the sun at 42° above the horizon.

FIG. 7 | EFFICIENCIES

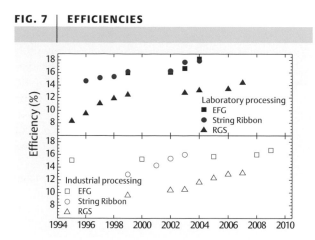

The evolution over time of the maximum efficiencies of solar cells made of ribbon silicon. The efficiencies of solar cells made by industrial fabrication methods in large-area quantities (below) track – with a certain time lag – the increase in efficiencies of small-area laboratory cells (above).

are first tested on a laboratory scale and then the attempt is made to transfer them to an industrial-scale production environment. In the case of string-ribbon and EFG silicon, efficiencies have already been achieved in this way which are not too far from those of solar cells made from poly-Si wafers fabricated by conventional technology. The processes are in both cases of similar complexity and thus have similar costs.

The application of ribbon-silicon wafers therefore makes a significant reduction in W_p costs possible due to its thrifty use of the expensive silicon starting material. Still greater cost savings should be obtained with solar cells based on RGS silicon wafers. However, their currently low efficiency must be further improved. This requires optimization of materials quality and of the fabrication processes. If it proves possible to forge ahead to efficiencies similar to those obtainable with the other types of silicon ribbons, the W_p costs could be further markedly reduced. The advantage would then be that producers of crystalline-silicon solar cells could make use of this cost-favorable wafer material in a straightforward way in their existing processing lines, since the main production steps would remain nearly the same. This would greatly facilitate the introduction of solar cells based on ribbon-silicon material.

Currently, in Broek op Langedijk in the Netherlands, an RGS pilot plant is being set up which will deliver RGS wafers on a second by second basis. This production machine is designed to improve the materials quality and therefore the obtainable efficiency, since it works at thermal equilibrium, in contrast to the laboratory devices thus far used. Then, at the latest, ribbon silicon should make a decisive contribution to lowering the production costs of solar modules.

Summary

Photovoltaic power generation will in the foreseeable future mainly make use of crystalline silicon as basic material, with an increasing tendency towards less expensive polycrystalline wafers. At present, the wafers are still sawed out of large silicon ingots, which results in considerable material waste by 'pulverization'. This increases the proportion of the wafer cost in the overall cost of the modules by up to 33 %. Ribbon silicon, on the other hand, makes use of a different preparation technology to avoid this material loss, and so yields considerable cost savings. In terms of efficiency, solar cells made from silicon ribbon are already nearly competetive with conventional cells. A further advantage of ribbon silicon is that it can be integrated into existing production processes for solar cells based on crystalline silicon with a minimum of problems. However, the most attractive production processes are not yet mature on an industrial scale.

References

[1] Photon **2012**, *4*, 42.
[2] C. del Canizo *et al.*, Progr. Photovolt.: Res.-Appl. **2009**, *17* (3), 199.
[3] T.F. Ciszek, J. Crystal Growth **1994**, *66*, 655.
[4] F.V. Wald, in: Crystals: Growth, Properties and Applications 5 (Ed.: J. Grabmaier), Springer-Verlag, Berlin **1981**.
[5] G. Hahn *et al.*, Solar Energy Materials and Solar Cells **2002**, *72*, 453.
[6] G. Hahn and A. Schönecker, J. Physics: Condensed Matter **2004**, *16*, R1615.
[7] S. Seren *et al.*, Proc. 22nd EU PVSEC, Milan **2007**, 854.

About the Author

Giso Hahn studied physics from 1989 to 1995 at the University of Stuttgart and carried out his Diplom work at the Max-Planck-Institut for Metals Research there. He received his doctorate in 1999 from the University of Constance and was leader of the group for "New Crystalline Silicon Materials" there from 1997 on. He completed his Habilitation (Dr.sci.) at Constance in 2005, and has been leader of the Photovoltaics Division since 2006.

Address: *Prof. Dr. Giso Hahn, Chair for Applied Solid-State Physics, University of Constance, Jakob-Burckhardt-Str. 29, 78457 Konstanz, Germany. giso.hahn@uni-konstanz.de*

CIS Thin-film Solar Cells

Low-priced Modules for Solar Construction

BY NIKOLAUS MEYER

Current solar modules are made of crystalline silicon. Their fabrication is complex and consumes a large quantity of materials and energy. Scientists and engineers have developed an alternative: CIS solar modules consume less semiconductor material, reduce costs, and now attain fully competitive efficiencies. Commercial systems are already on the market.

The absorption of sunlight is the most important step in photovoltaic energy conversion: Only sunlight which has been absorbed can be transformed into electrical energy. Silicon is a semiconductor; it absorbs light only when the energy of the incident light quantum (photon) is equal to or greater than the band gap energy of the material. In that case, the energy of the photon can induce an electron to jump out of the valence band and into the higher-lying conduction band, across the "forbidden zone" or band gap, storing the energy of the photon. Since electrons in the conduction band are also mobile, they can move through the material and thus contribute to a flowing electric current (see also the infobox "Compact Fundamentals of Photovoltaics" on p. 39).

Crystalline silicon is however an *indirect* semiconductor, so that for this transition to take place, the momentum of the electron must also be changed by interaction with the thermally vibrating crystal lattice. Since a simultaneous change in the energy level and the momentum of an elec-

CIS solar modules can form elements of a façade similar to darkened glass. In the example shown, they perform the typical functions of a cool façade: They cover the thermal insulation and provide for runoff of rainwater to the ground. The modules are designed as cassettes which can be mounted in the façade (cassette façade).

Renewable Energy. Edited by R. Wengenmayr, Th. Bührke. Copyright © 2013 WILEY-VCH Verlag GmbH & Co. KGaA, Weinheim

tron is less probable, it requires a disc of crystalline silicon (wafer) of 0.2 mm (200 m) thickness in order to completely absorb the incident sunlight.

CIS – an Ideal Material

For photovoltaic applications, direct semiconductors are therefore more interesting; their valence electrons do not have to change their momenta in order to accept energy from a photon. This class of semiconductors includes the compound semiconductors of composition $Cu(In,Ga)(S,Se)_2$ (abbreviated CIS). As minerals, they are called chalcopyrites. In their crystal lattices, there are as many copper atoms as the sum of the indium and gallium atoms, and half as many as the sum of the sulfur and selenium atoms. By varying the In/Ga and the S/Se ratios, one can continuously adjust the band-gap energy between 1.1 eV (for $CuInSe_2$) and 2.5 eV (for $CuGaS_2$), and thus the energy jump required of an electron which has absorbed a photon (1 eV = $1.602 \cdot 10^{(19}$ Joule).

Even as a very thin film of about 1 μm (millionth of a meter) thickness, CIS can almost completely absorb the incident sunlight. In comparison to conventional silicon solar cells, the amount of material required is thus reduced by a factor of 100. If the band gap of the CIS is adjusted to be near 1.5 eV, the material corresponds most closely to the solar spectrum, so that a maximum fraction of the sunlight can contribute to photovoltaic energy conversion. Its absorption coefficient and its band gap thus make CIS an ideal absorber material for the fabrication of solar cells.

Once the negatively-charged electrons have been excited into the conduction band through light absorption, a corresponding positive charge is left in the valence band in the form of quasiparticles, termed "holes". The second step in photovoltaic energy conversion now consists of separating the (excited) electrons from the holes. This task is performed by a p-n junction (see also the infobox "Compact Fundamentals of Photovoltaics" on p. 39). It is fabricated by connecting the p-type conducting absorber to a second semiconductor layer which exhibits n-type conduction. In a p-n junction, an electric field develops, and it separates the electrons from the holes. In the case of a CIS solar cell, the n-type semiconductor zinc oxide (ZnO) is employed as the n-conducting layer of the p-n junction. As with the negative pole of a battery, the negative charges collect in the zinc oxide layer when the cell is exposed to light, and the positive charge carriers collect in the (p-type) absorber.

In order to use this stored energy, the charge carriers must enter and move through an external circuit. To allow this to occur, the p-n junction is fitted with electrical contacts or electrodes – in the case of the CIS/ZnO junction, this is accomplished by a rear contact of molybdenum metal and a front contact of highly conducting zinc oxide (n-ZnO) (Figure 1). In practice, it has been found however that good-quality solar cells are obtained from the Mo/CIS/ZnO/n-ZnO structure only if a very thin buffer layer of zinc sulfide, cadium sulfide, or another semiconduc-

FIG. 1 | A CIS THIN-FILM SOLAR CELL

Sunlight —
0.003 mm
In the CIS absorber, sunlight is converted into electrical energy.

Zinc oxide front electrode
CIS absorber
Molybdenum rear electrode
Glass substrate

The conversion of sunlight into electrical energy in a CIS thin-film solar cell.

tor is deposited into the contact boundary region between the CIS and the ZnO layers.

Glass Coating instead of Wafer Technology

Glass is the substrate material of a CIS solar cell, and coating the glass substrate makes this otherwise passive construction material into a solar module for generating electric current. Fabricating CIS solar cells from wafers is neither possible nor reasonable, due to the tiny thickness of the photovoltaic active layer, which is of the order of 1 μm. Instead, glass serves as a low-cost substrate material, and the other materials which make up the solar cell are deposited layer-by-layer onto the glass surface. All together, the polycrystalline layered structure has a thickness of about 3 μm – CIS technology is thus a thin-film process (Figure 2).

For glass coating, various methods can be used to deposit the thin layers onto the substrate. Frequently, cathodic atomization or "sputtering" is used. For this technique, the desired sample material in the form of a solid block is introduced into a high-vacuum system, where it serves as the cathode for an electric field into which a noble gas (usually argon) is passed. If the field strength is sufficient, the gas is ionized by the field, a plasma ignites (as in a fluorescent lighting tube), and the ionized gas atoms are accelerated towards the cathode. Their impacts on the sample "atomize" the cathode material (thus "sputtering"), and as with ther-

0.003 mm

Fig. 2 *An electron microscope image of the polycrystalline layers of a solar cell which have been deposited onto glass. The molybdenum rear contact is at the bottom, CIS in the center, and ZnO/n-ZnO at the top.*

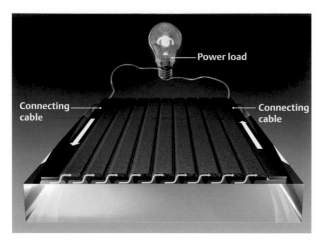

Fig. 3 *The electric current flow in a CIS solar module which has an integrated series circuit connecting the ribbon-shaped solar cells.*

Power load

Connecting cable

Connecting cable

mal evaporation, it enters the gas phase. This atomized material condenses onto the cool glass substrate surface opposite the cathode. This method yields layers of high quality which can cover areas of up to several square meters. Sputtering technology is used to apply the molybdenum and zinc oxide layers of a CIS solar cell.

The preparation of the CIS layer itself requires great care. Its composition, crystallinity and purity determine the quality of the solar cells. Most current technologies use either co-evaporation or a two-stage fabrication process. In the latter, precursor layers of copper, indium, and gallium are deposited onto the molybdenum-coated glass substrate – for this purpose, sputtering can be employed. This step determines the thickness of the layer and the composition of the final CIS film. In the second step, the precursor layers are heated in an atmosphere of sulfur or selenium-containing noble gas to a temperature of around 500° C. The gas phase reacts with the precursor layers to form the CIS compound.

Fig. 4 *A CIS solar module from the Soltecture company.*

In co-evaporation, all the required elements are evaporated by heating in a vacuum chamber at the same time. The CIS layer then grows layer by atomic layer on the glass substrate. Its composition can be precisely controlled and it can even be varied during the deposition process. This method yields CIS solar cells with the highest efficiencies.

A single CIS solar cell can currently generate a power output of about 20 mW/cm^2 when irradiated by the midday sun. A solar cell of 1 m^2 area would produce a current of up to 350 A at a voltage of 0.6 V. This high current would cause large power losses in the current-carrying resistive elements of the solar cell. For this reason, for a large-area photovoltaic element, the overall area is divided into small individual solar cells which are connected in series to form a complete module. In contrast to the case of silicon technology, this series circuitry can be integrated into the fab-

rication process: After deposition of the rear electrode, deposition of the buffer layer, and preparation of the front electrode, slightly shifted lines are scribed into the layers. This yields ribbon-shaped solar cells of 4 to 8 mm width, which are connected via a narrow contact from their front electrodes to the rear electrodes of the neighboring cells, and therefore function like batteries connected in series (Figure 3).

In order to protect the active layers of the solar module from weathering, they must be encapsulated. For CIS modules, this is usually accomplished by adding a second glass sheet which is laminated onto the coated substrate layer. As in silicon technology, suitable transparent glues are available for this lamination process. A black frame yields a glass solar module with a neat black appearance (Figure 4).

Ten Years of Industrial Experience

In the 1970's, scientists discovered the potential usefulness of CIS for making photovoltaic elements and fabricated the first solar cells from mm-thick single crystals of the material. An important benchmark was attained in 1980: Developers at the US firm *Boeing* succeeded in depositing polycrystalline layers of CuInSe$_2$ by co-evaporation, and used it to make a solar cell with over 10 % efficiency [1] (see also the infobox "Determining the Efficiency" on p. 50). Within a few years, the efficiency had been increased to 12.5 %, and the value of 15 %, which silicon solar cells were able to attain at that time, was within reach. Today, 30 years later, the best efficiencies of polycrystalline silicon and CIS solar cells are about the same, each slightly over 20 % [2].

The industrial potential of CIS technology was recognized early on: A CIS solar cell uses 99 % less expensive semiconductor material and two-thirds less energy for its production than a silicon cell. The technology thus plays a key role in reducing the costs of photovoltaic energy production.

The first commercial organization to enter the field was the petroleum firm *Arco*, which in the late 1980's began scaling up CIS technology and applying industrial production methods. It however took another ten years until their research group could release the first product for marketing, in 1998. Shortly thereafter, two other pioneers in the field, *Würth Solar* and *Soltecture* (at that time called *Sulfurcell*) began marketing solar modules, in 2003 and 2005, respectively.

Altogether, there are now more than twenty industrial suppliers active in the market today. None of them produces the same quantities as their competitors who manufacture solar cells from crystalline silicon. However, a number of these manufacturers have mastered CIS technology in production: Throughput, process yield and process stability today attain levels which make an expansion to gi-

FIG. 5 | INCREASING THE EFFICIENCY

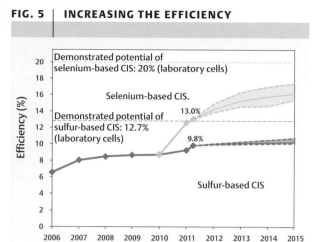

The continuous improvement of the efficiencies of commercial thin-film solar modules based on various CIS semiconductors, showing as an example results from the Soltecture company in Berlin. Blue-green: Cells based on sulfur-containing CIS, $Cu(InGa)S_2$; light green: Selenium-based CIS, $Cu(In,Ga)Se_2$. The latter exhibit a clear-cut jump in efficiency.

gawatt production capacities, comparable to the silicon-cell market, appear promising, with low risks and predictable economic rewards. The Japanese firm *Solar Frontier* has taken the lead in developing a correspondingly large production capacity.

It took thirty years of research and development to bring the CIS technology to maturity on an industrial scale. CIS represents a new class of semiconductors, about which much less is known than about silicon, which has been under investigation for a century and is generally one of the best-studied of all materials. Many of the materials properties of CIS were unknown for a long time, and some still are. The properties of large-area films had to be determined, the manifold influences of the fabrication processes on CIS quality required investigation, and the long-term behavior of the CIS materials in photovoltaic applications had to be studied. This continuously growing body of knowledge has been applied to increase the efficiencies of CIS solar modules further and further.

The example of our Berlin firm *Soltecture* makes it clear how the photovoltaic efficiency of a CIS compound could be continuously improved, and even – by modification of its chemical composition – given a quick boost (Figure 5). Selenium-based compounds of composition $Cu(In,Ga)Se_2$, in a laboratory format today already attain efficiencies of over 20 % [2]. Leading CIS manufacturers such as *Soltecture* currently obtain efficiencies of over 13 % for large-area module formats.

Solar Architecture using CIS Solar Modules

CIS technology is gaining importance not only because of its potential for reducing production costs; the solar modules are also attractive to architects and homeowners be-

cause of their impressive and attractive appearance. Their uniform anthracite-colored surfaces set them apart from solar modules made of polycrystalline silicon wafers, whose appearance is dominated by blue highlights and metallic conductor channels.

CIS technology is breaking the trail towards solar construction and promises to provide multifunctional façade materials for buildings. CIS modules can be used as construction materials for façades and roofs to protect against the weather, to contribute to thermal insulation, and to provide an architecturally aesthetic appearance, as well as to generate the electrical power required by the users of the building. Since the solar modules fulfill structural functions, the cost of passive structural materials can be saved. At the same time, energy production at the site, i.e. directly where the consumers are located, avoids costly and complicated transmission of electrical energy. Photovoltaics can be the key to decentralized electrical energy supplies with sustainable energy sources, and will be an indispensible component of the future energy mix.

Summary

Present-day solar modules are made of crystalline silicon. Their fabrication demands an elaborate process technology and consumes a large quantity of materials and energy. Scientists and engineers have developed an alternative, which is now in production on an industrial scale: CIS solar modules consume a much smaller amount of semiconductor material and reduce costs. At present, they attain competitive efficiencies of up to 13 % for commercial modules, and 20 % in the laboratory. Owing to their uniform black appearance, they are particularly attractive as components for solar architecture.

References

[1] R.A. Mickelsen and W.S. Chen, Polycrystalline Thin-film $CuInSe_2$ Solar Cells, in Proc. 16th IEEE Photovoltaics Specialists Conference (1982), 781–785.

[2] P. Jackson *et al.*, Prog. Photovolt. Res. Appl. **2011**, Wiley Online Library, DOI: 10.1002/pip1078.

About the Author

Nikolaus Meyer, together with Ilka Luck, founded the firm Sulfurcell Solartechnik GmbH in 2001 as a spinoff from the Hahn-Meitner Institute (HMI), and he is currently its managing director. He studied physics at the University of Hamburg and the TU Berlin, and managerial economics at the Hagen University. He obtained his doctorate at the FU Berlin for work on thin-film modules performed at the HMI, and worked at Siemens AG on their industrial fabrication. The founding of Sulfurcell was awarded a prize in 2001 by the Business Plan Competition of Berlin-Brandenburg.

Address:
Dr. Nikolaus Meyer, Soltecture GmbH, Gross-Berliner Damm 149, 12489 Berlin, Germany.
info@sulfurcell.de

CdTe Thin-Film Solar Cells

On the Path towards Power-Grid Parity

BY MICHAEL HARR | DIETER BONNET | KARL-HEINZ FISCHER

Thin-film solar modules made of cadmium telluride hold the promise of attaining so-called 'power-grid parity', as the first photovoltaic technology: The generation of solar power at prices which are competitive with conventionally-generated electric power.

Thin-film solar modules based on cadmium telluride (CdTe) [1] at present attain efficiencies of 11–12 % as commercial modules, and up to over 17 % with laboratory cells; theoretically, they could go as high as 30 %, exceeding the best theoretical efficiency of crystalline silicon. These CdTe solar modules thus have the potential of being the first photo-voltaic technology to attain so-called 'power-grid parity': They would then generate solar power at a cost which would be similar to that of power from conventional generating methods. This would make photovoltaics fully competitive.

The fundamentals of this technology have long been known. As early as 1970 in Frankfurt am Main, Germany, Dieter Bonnet, one of the present authors, constructed the first functional CdTe thin-film solar cell. CdTe solar cells have a number of attractive characteristics and thus an enormous commercial potential.

In contrast to silicon, CdTe is a direct semiconductor (see "Compact Fundamentals of Photovoltaics" on p. 39), like CIS, which was introduced in the previous chapter by Nikolaus Meyer. Thus, only a very thin layer of CdTe is required to completely absorb the incident sunlight. About a decade ago, the *ANTEC Solar* company, founded by Karl-Heinz Fischer and Michael Harr, manufactured the first commercial CdTe solar cells. These, however, still had absorber-layer thicknesses of up to 10 μm and therefore consumed relatively large amounts of CdTe. This was due to the still not completely mature technology for depositing the CdTe onto glass substrates at that time. Today's solar cells have CdTe layers which are only half as thick, with a tendency towards decreasing down to thicknesses of only 1 μm. This implies materials savings of 90 % in comparison to those first solar cells.

Furthermore, CdTe is an extremely stable chemical compound. In order to split a CdTe molecule into its component elements, cadmium and tellurium, one needs very high temperatures of 1000° C, or strong acids. CdTe is therefore stable over long time periods, an important precondition for attaining grid parity. The cost of a kilowatt hour of solar energy depends directly on the lifetime of the solar cells. In particular, CdTe does not degrade under the action of sunlight, so that the power output of the cells remains constant over many years. Their lifetime is limited not by the CdTe absorber, but rather by other components of the solar cells.

A further advantage of CdTe is its low temperature coefficient. This describes how strongly the efficiency of a solar cell decreases with increasing temperature: The larger the coefficient, the greater the decrease in efficiency. Since solar cells are heated by the sunlight, their efficiencies and thus their output power can decrease noticeably; this effect, however, is practically negligible for CdTe, and thus solar modules fabricated from it are suitable for use in regions with hot climates.

The efficiency of a solar module also depends upon the intensity of the sunlight. In most cases, it decreases when the insolation is lower, for example because of clouds or illumination at a grazing incidence angle. The fact that a photovoltaic installation using silicon cells yields less power on a cloudy day is thus due not only to the lower level of sunlight available for power generation; the less intense light is also converted to power with a poorer efficiency. CdTe, in contrast, shows very good behavior at low-level insolation, and makes better use of weak sunlight. CdTe solar power installations are therefore especially well-suited for use at higher latitudes where the sun is lower in the sky and there are often clouds.

The best Energy Balance and Lowest Costs

Of particular importance is also the short energy payback time for CdTe solar cells. This denotes – like the harvest factor referred to in other chapters on photovoltaics – the time during which a solar module must be operated under the local insolation conditions until it has yielded as much energy as was required for its manufacture. It thus depends on the site at which the module is operated. For CdTe solar installations, this payback time for use in Central Europe is less than 10 months. This value already includes the energy expended for precursor products, such as melting the sub-

Renewable Energy. Edited by R. Wengenmayr, Th. Bührke. Copyright © 2013 WILEY-VCH Verlag GmbH & Co. KGaA, Weinheim

strate glass, mining and processing the raw materials, etc. In Southern Europe or other regions with a high insolation, the energy payback time is even shorter. CdTe is thus the optimal technology in terms of its energy balance, also.

CdTe technology has the lowest manufacturing costs of all the currently-known photovoltaic technologies. For the future, independent studies predict that cost reductions are indeed to be expected for all the photovoltaic technologies, but CdTe will continue to maintain its lead [2]. Thus, the leading CdTe manufacturers today can fabricate solar modules at a cost of 0.5 €/W_P; in the future, a cost of less than 0.4 €/W_P is predicted. For other thin-film technologies, such as a-Si or CI(G)S, costs between 0.4 and 0.8 €/W_P are expected. A cost of around 1.2 €/W_P is predicted for the classical crystalline Si modules. Among all the photovoltaic technologies, CdTe thus offers the best perspective for attaining grid parity and being able to dispense with feed-in subsidies.

With this background, it is not surprising that for example the US firm *First Solar* (with European headquarters in Mainz) has expanded very rapidly. It produces exclusively CdTe solar modules, and within only a few years, it grew from a start-up to become the worldwide number one among solar module manufacturers. In the meantime, it has produced modules with an output power totalling 4 GW [3], equivalent to four nuclear power plant blocks.

Perhaps the ease of fabricating high-capacity solar modules from CdTe has however caused this technology to be almost completely overlooked by German research funding agencies. While classical silicon solar cells, thanks to intensive research support, today achieve efficiencies near 20 % and thus are approaching the theoretical limit for silicon, the efficiencies of 11–12 % attained by CdTe modules are only slightly more than 1/3 of the theoretical value. Here, there is still much room at the top. A concerted research effort could still further reduce the manufacturing costs quoted above, perhaps even halve them. Countries like China, which have realized the potential of this technology, are therefore investing not only in CdTe solar-module factories, but also in CdTe research.

A simple Coating Procedure

One of the reasons for the favorable fabrication costs of CdTe solar cells is an important property of CdTe: it sublimes congruently. 'Sublimation' refers to the vaporization of a solid material, without its previously having melted. Due to sublimation, snow 'vanishes' in dry winters; the water molecules go directly from the solid ice into the air, as water vapor. CdTe exhibits a similar behavior: When it is heated to a temperature above about 500° C, CdTe vapor is formed directly. If a sheet of glass at a lower temperature is held in this vapor, the latter condenses as a thin film on the glass.

Now the property 'congruent' enters the picture: It means that the CdTe vapor maintains precisely the same chemical composition as in the solid CdTe from which it

FIG. 1 | THE CSS PROCESS

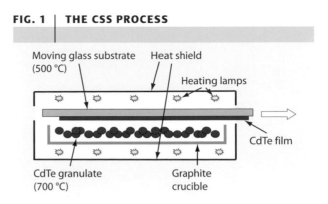

The principle of the close-spaced sublimation process.

originates. Thus the deposited layer has exactly the same composition as the solid source material, and therefore as the pure starting material. In this respect, CdTe differs essentially from the class of CIS and CIGS materials (see the previous chapter). In that materials class, the ratio of C (copper) to In (indium) atoms is of decisive importance for their photovoltaic efficiency. However, if one simply vaporizes these compounds, the vapor would have a different C/In ratio from the starting material, and the deposited layer would have still another ratio.

During fabrication on an industrial scale, the composition of the deposited layers thus threatens over the course of time to deviate from that of the starting material. Furthermore, such a system tends to show local variations in composition over the surface of a large-area deposited layer. Therefore, solar-module production using CIS and CIGS demands sophisticated deposition processes.

For CdTe, in contrast, the deposition process is extremely simple and robust (Figure 1). A graphite crucible is filled with CdTe and heated in vacuum to 600–700° C. The glass substrate sheets are passed over the crucible at a distance of a few millimeters, after being preheated to a temperature of 500–600° C. The subliming CdTe has only a short path to the glass substrate, where it condenses as a thin, dense and mirror-smooth film. One need take no special care to maintain the composition of the deposited layer; the physics and chemistry of CdTe ensure that it will condense at the ideal composition. This method is called the CSS process (for close spaced sublimation), since the sublimation and condensation are spatially close together.

A thin CdTe layer on glass of course does not by itself represent a functional solar cell. In addition to a transparent front contact and a rear contact to carry off the current generated in the cell, a p-n junction must be produced. There, the charge carriers generated in the CdTe absorber are separated. To make this junction, a very thin (~100 nm) layer of CdS is deposited on top of the CdTe layer, likewise using the CSS process (Figure 2). CdS condenses automatically as an n-type semiconductor, while the CdTe film is a p-type semiconductor, so that the desired p-n junction results. A theory of the mode of action of the CdS layer was

FIG. 2 | THE STRUCTURE OF A SOLAR CELL

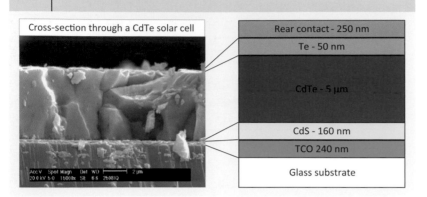

Cross-section through a CdTe solar cell

Rear contact - 250 nm
Te - 50 nm
CdTe - 5 μm
CdS - 160 nm
TCO 240 nm
Glass substrate

This electron microscope image shows a cross-section through the layered structure of a typical CdTe solar cell. The sunlight is incident from below through the glass substrate and onto the absorber layer.

given by Karl W. Böer [4], among others. In order to achieve high efficiencies, the CdS/CdTe junction must however be subjected to a special thermal treatment after deposition. This removes defects at the boundary between the two layers, at which the charge carriers generated would otherwise recombine and be lost.

A new Coating Process

In order to produce high-quality films which yield high efficiencies, two things are necessary: A high substrate temperature, over 500° C, and precisely the short source-substrate distance described above. In general, a simple rule of thumb holds: the higher the substrate temperature, the better the quality of the deposited layer. This however presents a technical challenge. In order to fabricate solar modules at a moderate price, common window glass is used as the substrate; but it begins to soften and flow at a temperature just above 500° C.

For this reason, sheets of window glass with a large surface area can be coated using the CSS process only up to about 520° C. Several manufacturers of CdTe solar modules however work with higher substrate temperatures. They therefore support the softening glass with a number of rollers below it, and deposit the film from above. But this means that the distance from source to substrate over which the CdTe molecules must travel becomes much longer. Without special preventative measures, the molecules would collide with each other underway and would stick together; somewhat overstated, they would agglomerate to dust. The resulting films are less dense and less smooth, and tend to form pinholes. Without further treatment, such layers as a rule have only very low photoefficiencies.

To overcome this problem, we developed a special configuration for the crucibles and the rollers at our firm *CTF Solar* in Kelkheim. It allows us to pass glass substrate sheets closely above the hot crucible even at temperatures higher than their softening temperatures without notable plas-

tic deformation, and thus to deposit high-quality CdTe films over a large surface area.

With a team of around twenty physicists, chemists and engineers, we design and build highly productive CdTe solar-module factories. In the investment costs for the construction and commissioning of such a factory, an important factor is the cost of the machines required. A factor which is often neglected, however, is the cost and the time needed for the ramp-up of a factory. This is the pilot phase of production, when the efficiency of the solar modules produced and the productivity of the factory are optimized to their target values. The task to be carried out is illustrated by Figure 3, which shows the time variation of the efficiency of the solar modules during a production run. The figure shows an example of production during the early ramp-up phase. On the one hand, it can be seen that the statistical scatter in the values of the efficiency is extremely low, as expected for the deposition of CdTe using the CSS process. On the other hand, the figure shows clearly that the production process is not yet stable. The efficiencies exhibit jumps and long-term drifts. Both of these effects are due to instabilities in the process machinery.

One of the principal tasks during ramp-up is to eliminate these instabilities as quickly as possible. This can be done only in cooperation with the manufacturer of the machines. We therefore prefer to work with machine manufacturers who are willing to consider the needs of the customer already in the design phase of the machines; they are then also involved in the fine-tuning of the machines in the customer's factory.

Secondly, a rapid ramp-up requires sophisticated diagnostics. With the corresponding diagnostic tools, after each

FIG. 3 | RAMP-UP

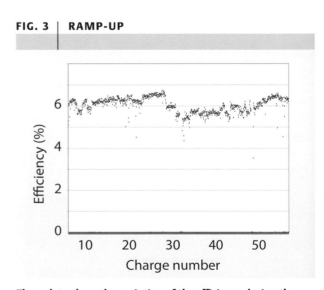

These data show the variation of the efficiency during the initial phase of ramp-up. At this point, the average efficiency is only around 6 %. This level was reached after fabrication of about 2,000 solar modules. The small statistical variation, typical of CdTe, is notable. It is however superimposed on systematic jumps and drifts, which are caused by the process machinery.

important process step the results are checked 'in-line', i.e. on every single substrate sheet, and are analyzed in terms of deviations from the desired process characteristics. At the same time, the state of each of the machines is registered by a large number of sensors. A manufacturing execution system (MES) combines these two data sets and analyzes them. This quickly reveals which of the machine parameters have a negative influence on the efficiencies and require correction.

Figure 4 shows a model of a production line with a production capacity of up to 500,000 solar modules per year. This plant, which measures 180×70 m, produces modules that are 1.2 × 1.6 m in size; they are thus more than 2.5 times larger than the standard dimensions of 0.6×1.2 m. The annual production thus consists of nearly 1 million m². With a module efficiency of 11 %, which can be achieved after an intensive ramp-up, this corresponds to an output power of over 100 MW$_P$.

Delivery of such a production line as a turn-key product requires an investment of 80 to 100 million €, depending on the extent of the ramp-up. This price may seem high, but it should be seen in connection with the production capacity of 100 MW$_P$ annually: The specific investment sum is less than 1 €/W$_P$, and is thus even lower than for other thin-film technologies. Such a turn-key line is so to speak a starter set for establishing a solar-module factory. When the production capacity is increased by adding other similar production lines, the effort required for ramp-up decreases and thereby also the specific investment costs, down to the range of 0.5–0.6 €/W$_P$.

Summary

Thin-film solar modules made from cadmium telluride (CdTe) can already be fabricated at moderate cost with efficiencies of 11–12 %. This cost advantage of CdTe relative to other photovoltaic technologies will persist in the future. Their market share as thin-film solar modules grew continuously in the past years. They are developing into the 'workhorse' of the solar technologies. In addition, they will achieve power-grid parity as the first of the photovoltaic technologies, and thus will deliver electric power at competitive prices. More intensive research should allow their efficiencies to increase by a factor of two and also lower their production costs notably.

References and Links

[1] D. Bonnet and M. Harr: 2nd World Conference & Exhibition on Photovoltaic Solar Energy Conversion, Vienna **1998**.

[2] U.S. Department of Energy: Solar Energy Technologies Program, 2007–2011.

[3] www.firstsolar.com.

[4] K.W. Böer, Solar Cells **2010**, 95, 786.

FIG. 4 | A CdTe SOLAR-MODULE FACTORY

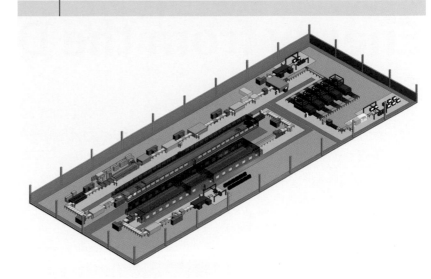

A production line for CdTe solar modules, with an annual production capacity of up to 100 MW$_P$.

About the Authors

Michael Harr, Dieter Bonnet, Karlheiz-Fischer (f. l. t. r.).

Michael Harr *studied physics and business administration at the University of Göttingen, where he obtained his doctorate in 1977. In 1993, along with Dieter Bonnet and others, he founded the technology concern ANTEC, and in 1997, together with Karl-Heinz Fischer, the firms ANTEC Solar and ANTEC Technology. He was Shareholder-Managing Director and Technical Manager of these three companies. Together with Dieter Bonnet and Karl-Heinz Fischer, he built the world's first commercial solar-module factory producing CdTe thin-film modules for ANTEC Solar, from 1998-2001 in Arnstedt (Thüringen). In 2007, he initiated the founding of CTF Solar, which he manages at present.*
Dieter Bonnet *developed the first CdTe thin-film solar cells as a pioneer in the early 1970's. In the technology concern ANTEC, he managed the Solar Energy sector and developed fabrication processes for CdTe modules. Since retiring, he is a consultant for CTF Solar. In 2006, the Commission of the European Union conferred on him the Becquerel Prize for his work in photovoltaics.*
Karl-Heinz Fischer *was a founding member of the solar companies ANTEC Solar and ANTEC Technology, and was Shareholder-Managing Director of those firms. As a member of the board of directors of the Federal Association for Solar Economy, he played a major role in establishing the Renewable Energy Act (EEG). Currently, he is sales manager of CTF Solar.*

Address: *Dr. Michael Harr, Managing Director, CTF Solar GmbH, Industriestr. 2–4, 65779 Kelkheim, Germany, m.harr@ctf-solar.com www.ctf-solar.com.*

Geothermal Heat and Power Generation

Energy from the Depths

BY ERNST HUENGES

The earth can provide much heat energy to drive geothermal base-load power plants – everywhere, not just in active volcanic areas. However, tapping this energy requires new exploitation technologies.

The Geothermal Laboratory in Gross Schönebeck, Germany.

Ambitious energy and environmental policy goals are creating new challenges for energy suppliers. The energy mix of the future will have to be ecologically friendly, secure in resources, and competitive. Long-term security of the energy supply and most especially sustainability are in demand. The goal of the European Union to increase the contribution of energy from sustainable sources to 20 % of the total used by the year 2020 emphasizes the high expectations which are placed in these energy carriers.

The geothermal potential is extremely interesting in the context of environmental and economic policy. In contrast to wind und solar energy, it is available around the clock, which makes it attractive for base-load power plants. Even taking into account the energy expended in constructing the installations, geothermal energy produces only a small amount of CO_2. It represents an ecologically exemplary and expedient alternative to nuclear power and to fossil fuels.

The earth contains a high potential for supplying heat for the energy economy. Its heat content results from the release of gravitational energy by the contraction of gas and solid particles during its formation, as well as from primordial heat dating from the early days of the solar system, and from the energy released by the decay of radioactive isotopes. According to current knowledge, the isotopes which are significant for heat production are those of uranium, thorium and potassium that are enriched in the continental crust, which consists mainly of granitic and basaltic rock (see the infobox "Heat from within the Earth" on p. 63).

Geothermal heat sources can supply energy in the form of technologically usable heat or electric power. The occurrence of geothermal heat sources is not limited to regions with noticeable vulcanism. In principle, there is geothermal heat everywhere, in particular also under Central Europe (Figure 1). In Germany, three regions are suitable for extracting deep geothermal energy for electric power generation: the South German Molasse Basin, the upper Rhine Graben, and the North German Basin. To be sure, there one must drill down to depths of several kilometers in order to tap a level of temperatures which is high enough to effectively generate electric power using steam turbines. Real-

izing this potential makes special demands on technology and engineering, and at the present state of development, it involves high investment costs. Technologically and economically viable concepts are needed in order to increase the small fraction of the energy market which is currently supplied by geothermal resources.

Geothermal Energy is Still Exotic, but has High Growth Rates

Worldwide, in 2011 roughly 11 GW of electric power was generated using geothermal energy [1]. For comparison: In the same year, the total installed power output of all the world's wind power plants was nearly 160 GW [2]. However, wind power depends on the weather, so that all the wind power plants on the globe practically never produce their maximum power at the same time. In the case of geothermal power, in contrast, this would be possible, at least in principle.

Unlike Iceland, in continental Europe, geothermal energy plays only a subordinate role. Only Italy can claim a significant number of geothermal power plants, which together produced more than 800 MW of electric power in 2011 [1]. Larderello, Italy, is also the birthplace of the extraction of electrical energy from geothermal heat: In 1904, Count Piero Ginori Conti installed a dynamo there which was driven by steam from the volcanic terrain. It lit up five incandescent lamps in the village.

In Germany, the utilization of geothermal heat has seen relatively high growth rates in recent years. At the end of

Renewable Energy. Edited by R. Wengenmayr, Th. Bührke. Copyright © 2013 WILEY-VCH Verlag GmbH & Co. KGaA, Weinheim

2011, over a GW of heat production capacity from geothermal resources was installed. Of this, 160 MW$_{th}$ was supplied to larger plants, while more than one GW$_{th}$ was additionally obtained from geothermal probes: These are heat sources for heat pumps which typically are used to heat single- or multiple-family dwellings [3]. In southern Germany, hot deep-well water is used in several towns for central heating, for example in Erding, Pullach, and Unterschleißheim. Geothermal installations in Mecklenburg-Vorpommern have contributed to fulfilling space heating needs since the 1990's.

But the tapping of deep geothermal reservoirs for generating electric power also shows a positive trend. In November, 2003, in Neustadt-Glewe in the state of Mecklenburg, Germany's first geothermal pilot power plant began operation, with an output power of 0.2 MW$_{el}$. It demonstrated the technical feasibility of generating electric power with geothermal energy under the conditions prevailing in Germany. In Unterhaching, Bavaria, a binary plant, i.e. power generation using a secondary cycle with a heat exchanger heated by thermal water from underground, has supplied the town with geothermal power since 2009, after it began delivering heat to the district heating network in 2007. This particular plant in Unterhaching has a so-

called Kalina cycle. In the secondary cycle of the Kalina system, a mixture of water and ammonia is heated, producing gas at relatively low temperatures. The hot gas drives a turbine, which is connected to a generator to produce electric power. This circuit thus works with a "temperature head" that is not sufficient to be used with conventional steam turbines. Another common binary process is the Organic Rankine Cycle (ORC). Here, an organic working fluid is used to drive the turbines (Figure 2). In Landau, the first large, industrial-scale geothermal power plant in Germany went on line in 2007, providing 3 MW$_{el}$ of electrical output power and a heat output of 8 MW$_{th}$ for a district heating network. Figure 3 gives an overview of German geothermal heat supply projects; additional plants are in the planning and construction stages.

According to the Geothermal Report which was adopted in May 2009 by the German Federal cabinet, the goal is to have an installed electric generating capacity of 280 MW from geothermal sources operating by the year 2020 [5]. At an output power of ca. 5 MW per plant, this corresponds to more than 50 plants. Together, they can potentially provide about 1.8 TWh of electrical energy and additionally 3.4 TWh of heat each year. Regarding heat supplies without electric power generation, it is expected that by 2020,

FIG. 1 | TEMPERATURES IN THE DEPTHS OF THE EARTH

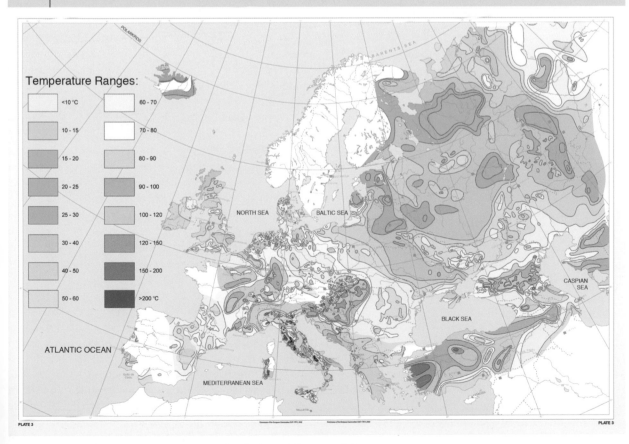

The temperature distribution in Europe at a depth of two kilometers [6].

an additional 4.8 TWh can be supplied from geothermal energy. The total heat supply from geothermal energy would then be 8.2 TWh per year.

For comparison: Germany's consumption of light heating oil between 2004 and 2006 amounted to more than 1000 pctajoules per year; this corresponds to ca. 270 TWh [4]. After 2020, the growth of geothermal energy supplies is expected to accelerate, leading to an installed electric generating capacity of 850 MW by 2030 [5].

Geothermal Energy Sources

Relatively widespread in Central Europe are near-surface geothermal resources: Heat pumps use surface and ground water from a depth of a few meters as a heat source for space heating in houses. Installations of this type require only a few degrees of temperature difference in order to yield sufficient heat. A second heat source is hot water from deeper within the earth. Such hydrothermal systems can be found in areas with active vulcanism, but also in non-volcanic regions. Today, most of the large geothermal power plants in the world use hot water from volcanically active regions to generate electric power.

Hydrothermal systems which are not directly connected with a volcano cause fewer technical problems. In Southern Germany and on the North German Plain, there are for example several regions with hydrothermal low-pressure reservoirs at depths up to about 3000 meters. These areas of hydrothermal potential are aquifers which carry hot water – usually salty – that can be brought to the surface through wells.

Since this water is at temperatures between 60 and 120° C, it is hardly suitable for effective electric-power generation. Therefore, it is mainly used for space heating. The heat from the deep-well water is transferred in heat exchangers

FIG. 2 | GEOTHERMAL POWER GENERATION: THE ORC PROCESS

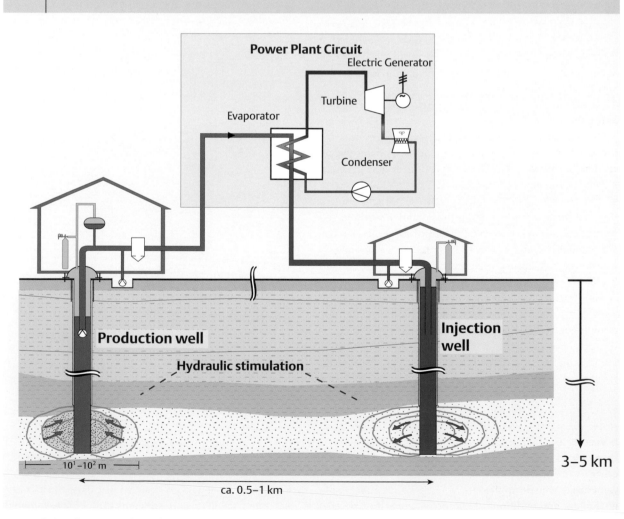

A pump brings hot water through a production well from deep within the earth to the surface. Its heat is used in a vapor generator to drive a turbine for electric power generation within a power plant circuit. The turbine circuit contains an organic working fluid with a low boiling point in order to increase its efficiency (Organic Rankine Cycle). The cooled water is pumped back to the depths through an injection well (blue) (graphics: GFZ).

to the district heating grid. In this range of temperatures, there are a number of possibilities for utilizing the geothermal heat – apart from electric power generation. Typical examples are central heating installations for local and district household heating, small consumers and industrial applications, which represent the state of current technology. The direct use of thermal water for bathing and in therapeutic baths is a classic example.

Hot and Deep

Below a depth of 4,000 meters, one finds rock formations at temperatures above 150° C practically everywhere under the Earth's surface. They contain by far the largest reservoir of geothermal energy which is currently technically accessible and interesting in terms of electric power generation. The technology for utilizing deep-well geothermal energy usually requires two boreholes: one production well and one injection borehole, in order to obtain hot water from a deep-level reservoir (aquifer). Installations with a larger number of boreholes (geothermal fields) are not yet the state of the technology in Germany. Above ground, the thermal water circuit is closed. As a rule, the thermal energy is transferred via a heat exchanger to the user application (the steam cycle of a turbine, heating network, heat pump etc.). The cooled thermal water is then pumped back into the aquifer via the injection borehole.

A central role in the future usage of deep-well geothermal energy is played by Enhanced Geothermal Systems (EGS). Enhanced, or engineered, geothermal systems are those which have been made economically competitive by the use of productivity enhancing measures, as will be described in the next section. Among these EGS are the previously so-called Hot Dry Rock (HDR) systems. These are dry rock formations which are used with water injected from above ground. This water takes up heat from the rocks via naturally-existing or artificially-produced (by 'stimulation') heat-exchanger channels and carries it up to ground level.

EGS technologies were developed for sites which are initially not cost-effective. Around 95 % of the geothermal potential in Germany can be realized only with the aid of these technologies. All of the necessary system components are in principle available, but their combination as yet often does not work with sufficient reliability and efficiency.

Thus, for example, the sedimentary basin systems which are widespread in the North German Basin and contain hot water represent a promising potential for providing geothermal heat. Since such systems are widespread all over the world, the technologies developed here for utilizing geothermal energy are relevant not only for Germany. They can equally well be applied to other sites with comparable geological preconditions. This international perspective should provide an additional boost for the technological development.

Sedimentary geological regions are most often found in areas with a high population density and a potentially great

HEAT FROM WITHIN THE EARTH

The sun, to be sure, radiates 20,000 times more energy onto the earth's surface than it receives in the form of heat from the depths. Nevertheless, geo-thermal heat is a practically inexhaustible source of energy on a human scale. This heat energy originates from three sources [4]:

- The gravitational energy stored in the interior of the earth;
- The primordial heat energy stored in the earth's interior; and
- The decay of natural radioactive isotopes.

As the earth was formed from the protoplanetary nebula by accretion of matter, i.e. chunks of stone, dust, and gases, its mass increased and with it its gravitational field. Thus, the matter that continued to rain down on the nascent earth impacted with increasing force, and the gravitational energy released was converted for the most part into heat. A large portion of this heat was, to be sure, radiated back into space, but estimates show that an energy of between 15 and $35 \cdot 10^{30}$ J remained in the proto-earth. An additional quantity of energy came from the heat which the matter from the protoplanetary nebula itself brought to the nascent earth.

In the earth's continental crust, the decay of natural radioactive isotopes makes an important contribution to geothermal heat. Especially in the layers near the earth's surface, the naturally-occurring isotopes ^{40}K, ^{232}Th, ^{235}U, ^{238}U and others are enriched in the granitic and basaltic rocks. In the basaltic rocks, the radiogenic heat production leads to a power output of around $0.5 \ \mu W/m^3$; in granites of up to $2.5 \ \mu W/m^3$. Since the formation of the earth, this source of heat has released at least an estimated $7 \cdot 10^{30}$ J of energy.

According to modern estimates, these three sources lead to a total heat energy stored in the earth of between 12 and $24 \cdot 10^{30}$ J. The earth's outer crust down to a depth of 10 km thus contains about 10^{26} J. The resulting heat current towards the surface is around 65 mW/m². For each kilometer downwards into the earth, the temperature in the outer crust increases on average by 30° C.

(Roland Wengenmayr)

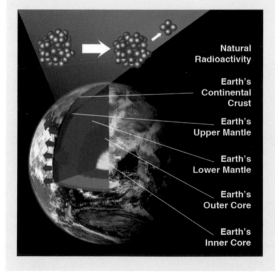

Natural Radioactivity

Earth's Continental Crust

Earth's Upper Mantle

Earth's Lower Mantle

Earth's Outer Core

Earth's Inner Core

In the continental crust, natural radio- activity produces strong geothermal heat *(graphics: Roland Wengenmayr).*

demand for heat, which can be met without much additional effort in conjunction with the construction of geothermal power plants. The fractured and porous stone which predominates in such regions serves as a heat and water storage reservoir; it can be regarded as a geothermal reservoir, however, only when it lies at considerable depth, due to the moderate heat flow from deeper in the earth. These zones contain deep, porous layers of carrier rock, which hold water (aquifers) and sufficient stored energy content to serve as a source for geothermal power generation. The transport paths for this water are either natural and occur as an open system of interfaces, or else must be artificially produced through stimulation.

FIG. 3 | DEEP-WELL GEOTHERMAL PROJECTS IN GERMANY

FIG. 4 | THE GROSS SCHÖNEBECK BOREHOLES

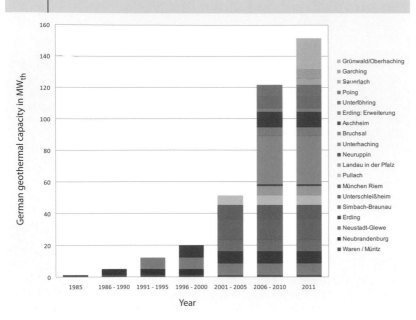

The heat-production capacity in megawatts (MW$_{th}$) of geothermal installations in Germany with boreholes more than 1000 meters deep (data as given by the operators).

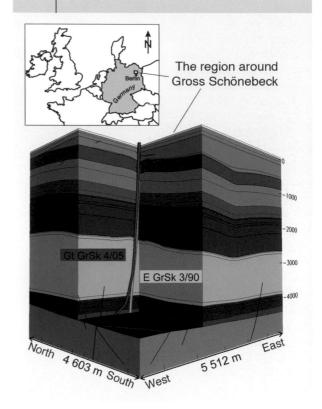

The paths of the boreholes and the geological profile around the Gross Schönebeck research area. Red represents the older borehole E GrSk 3/90, dating from a natural gas exploration in 1990. Blue indicates the geothermal borehole Gt GrSk 4/05 from 2007, which is 4,400 meters deep.

Reservoir Engineering

The exploitation of underground heat depends to a great degree on the efficient management of the reservoir. To this end, one must understand the whole system, consisting of the boreholes and the underground reservoir, qualitatively and quantitatively. This knowledge then allows the structuring of the processes which take place in the boreholes, in the nearby zones, and in the reservoir itself.

The principal method for increasing productivity is reservoir engineering. It permits the properties of deep-lying geothermal reservoir stone to be modified in a controlled way. If the natural porosity of the stone is limited, so that the water flow rate would be too low and the interface area too small for effective heat exchange, special stimulation methods can be applied to produce additional fractures in the rock. One method is hydraulic stimulation of the cleavages and fractures. Hydraulic fracturing is a well-established procedure in the petroleum and natural-gas industry. Developed in the 1940's and continuously improved since, it is employed there to increase the productivity of oil and gas wells. Hydraulic fracturing is increasingly playing a key role in the exploitation of geothermal heat, as well, since cost-effective conversion of geothermal heat into electric power requires not only a sufficiently high temperature as mentioned above, but also a stable supply of large quantities of thermal water.

In order to provide this supply, the rock layers must be highly porous and have a good permeability, i.e. a high proportion of hydrologically-connected pores (hot fractured rock). This permits a high rate of percolation and a good flow rate to the boreholes. However, at depths where the temperature is around 150° C, the natural permeability is usually limited. The stone must be artificially fractured in order to permit an unobstructed water circulation. In addition to producing a far-ranging system of fractures, connections to water-bearing cleavages are opened up. Stimulation creates the Enhanced Geothermal Systems already mentioned above.

In hydraulic stimulation, within a short time a fluid under high pressure, usually water, is injected through a borehole. The pressure of the fluid is greater than the stresses present in the rock formations, and thus it widens already existing fractures in the stone, connects them, and opens new cleavages. If necessary, the stimulation fluid is loaded with a support medium to secure the fractures being opened up, for example with small ceramic spherules of ca.

Fig. 5 *A nitrogen test in 2007 at the borehole 4/05 in Gross Schönebeck (photo: GFZ).*

1 mm diameter. These lodge in the fractures and keep them open when the pressure is released. Stimulation produces an extensive system of fractures that provides flow channels for the thermal water to the boreholes. They function both as transport paths and for heat exchange with a large interfacial area.

The Geothermal Laboratory in Gross Schönebeck

In order to develop practical approaches for new technological methods and to solve problems encountered in test operations, demonstration plants are indispensable. The Helmholtz Centre Potsdam – German Georesearch Centre (GFZ), within its research program "Geothermal Technologies", has set up a so-called *in-situ* geothermal research laboratory at Gross Schönebeck in the German state of Brandenburg. It is the only installation worldwide for investigating the exploitation of geothermal heat from large sedimentary rock formations under natural conditions. It carries out hydrologic experiments and borehole measurements which yield information about the geological and hydrogeological conditions at great depths. The experiments also give valuable indications of the behavior of the stone formations under the application of modern reservoir engineering methods such as hydraulic stimulation.

Two boreholes over 4 kilometers deep give access to interesting horizons in the North German Basin at depths between 3,900 and 4,300 meters and temperatures around 150° C. They serve as a natural laboratory for investigating geological and technical aspects of drilling deep boreholes. Since 2001, a number of successful series of hydrologic experiments and borehole measurements, with emphasis on controlled increase of productivity in geothermal repositories, have been performed (Figures 5 and 6).

The main goal of these investigations is the solution of problems to support long-term, site-independent usage of geothermal resources. The emphasis lies on developing technical measures for controlled stimulation of various rock formations, in order to be able to utilize geothermal heat everywhere where it is needed. The development, testing and optimization of specific methods and processes relevant to geothermal energy is intended to make possible the construction of cost-effective geothermal heat and power plants on a long-term basis by introducing innovative concepts. Reservoir engineering knowledge gained e.g. from stimulation experiments is needed wherever fluids are transported underground using wells. Lessons learned from geothermal projects are useful for all underground systems where fluids are to be extracted from deep underground, or materials are to be stored at depth. The results of these investigations are thus also relevant to the topic of CO_2 sequestration.

Along with stimulation procedures, new, innovative methods of opening up geothermal reservoirs and of directed drilling in the region of the reservoir are being applied at Gross Schönebeck (Figure 7). The drilling of geothermal boreholes presents a particular challenge. The planned utilization of a geothermal reservoir over a timespan of twenty to thirty years requires that the reservoir be opened with care and that secure drilling methods be used. The experience of the hydrocarbons industry (petroleum, natural gas), which is the leader in the area of drilling tech-

Fig. 6 *Thermal water brought up from the depths is collected and measured in containers. In later power-plant operation, the hot water produced and used will be returned underground through a second borehole (photo: GFZ).*

Fig. 7 *A rotary bit in action during the drilling in 2006/07 at Gross Schönebeck. Steerable drilling systems allow the borehole to be sunk in a precisely directed manner to kilometer depths along horizontal, vertical and inclined paths* (photo: GFZ).

nology, is taken into consideration. The potential profits from exploration and development of oil and gas fields has driven important technical innovations.

However, experience has shown that one cannot simply adopt the methods of the hydrocarbons industry without modification, as the requirements of geothermal applications are somewhat different. For example, new solutions for completion technologies as required specifically for geothermal applications must be developed and tested. 'Completion' refers to finishing off the borehole, including the installation of casing (Figure 8). Special goals for geothermal applications are minimal damage to the reservoir on opening it via boreholes, obtaining a maximal reservoir area through which thermal water can flow, and long-term protection of the boreholes in a corrosive environment. This includes among other things increasing the drilling velocity, applying measurement technology at depth to facilitate the use of directed drilling techniques, and monitoring for reliable operation.

Experience shows that opening of a geothermal reservoir will have a positive influence on its energy yield. This minimizes the high costs and risks of the exploitation of geothermal reservoirs and increases the output flow of thermal water. Geothermal repositories can thus be prepared for the desired long-term usage and their production of thermal water can be improved. In view of the increasing num-

bers of geothermal installations, there is a considerable potential for cost savings. The results of these investigations are significant not only for users and for planning and engineering agencies, but also in view of the increasing interest in investments in geothermal energy, for the initiation of geothermal drilling and reservoir projects.

An additional focus is on materials research. The high salinity of geothermal fluids can lead to an increased corrosion danger in the system components used and thus to massive problems for the operation of geothermal plants. In order to make the right choice of cost-effective materials suited to a particular site, investigations of materials quality are also being carried out at Gross Schönebeck. A specially-designed corrosion test installation for that purpose consists of a system of piping bypassing the above-ground thermal water circuit. In this system, the corrosion resistance of various metallic materials under realistic conditions is tested "*in situ*", on the basis of electrochemical measurements. A number of connecting valves allows the installation of different model components (piping), materials samples, sensors and also a heat exchanger with varying interface materials, thus giving information about the dynamics of corrosion processes in these subsystems.

The research power plant in Gross Schönebeck is completed with an ORC circuit that will permit the study of flexible operations procedures, e.g. for partial-load operation. The power plant cycle will be constructed in a modular design, with three circuits, separated according to their working media and containing preheaters, vapor generator, turbogenerator and condenser, among other components. These three modules will be charged one after the other with their thermal media, with each module designed for different process constraints. In this manner, efficient power generation can be demonstrated. Furthermore, different components and working media can be tested under varying simulated site conditions. The geothermal heat is transferred to the power plant module via an intermediate circuit. This has on the one hand the advantage that only one heat exchanger is required for the thermal water. On the other hand, all preheaters and steam generators make use of the same well-defined working fluid on their heat-input sides, which notably improves the quality of future component evaluations. Modifications of the intermediate circuit also permit variations of the hot-water temperature.

Utilization concepts applicable independently of the site will make possible a broader usage of geothermal resources even outside geothermal anomalies such as regions of volcanic activity. The results of investigations at the Gross Schönebeck reference site are thus a precondition for widespread industrial-scale exploitation of geothermal energy in the North German Basin and elsewhere.

Need for Further Research

In December, 2006, a seismic event of magnitude 3.4 on the Richter scale was registered in Basel, Switzerland. This event attracted public attention to deep-level geothermal energy and its risks. It is particularly important that the experience gained there be used constructively in future geothermal project planning. The earth's underground is complex and heterogenous. Every potential project site requires extensive preliminary geological investigations and a concept for its implementation which takes into account the particular characteristics of that site. Seismic monitoring deserves special attention; it forms the basis for estimates of the seismic risks and scenarios for minimizing them. These preliminary investigations should already form the basis for weighing the pros and cons of the project. In geothermally favorable regions such as the Upper Rhine Graben, the natural risk of seismic activity is generally higher than for example in the North German Sedimentary Basin, where seismic activity is hardly to be expected.

On the basis of well-founded research, the next step towards utilization of geothermal energy can and must be taken. Demonstration projects accompanied by scientific monitoring, which do not aim at rapid commercial success, are especially important. Research and development, pilot plants and demonstration installations have laid the groundwork in recent years for many successful geothermal projects in Germany. Deep-level geothermal energy is still however in its infancy, the technological challenges are great, and they will require time to be overcome. Successes and failures must be combined in a reasonable manner, data and experience must be exchanged and evaluated. Solutions to problems and best-practice scenarios must be developed on this basis and made available to future project planners.

The population needs to be involved in this process. The great opportunities of geothermal energy – but also its problems and uncertainties – have to be communicated to the public and understood. A successful development and wide application of geothermal energy sources will be possible only with the acceptance and support of the public on a broad scale.

Outlook

In Germany, geothermal energy today contributes only a minor amount to the overall energy supply; however, in recent years it has shown notable growth. The first installations for coupled heat-and-power production have begun operation. They demonstrate that geothermal resources represent an interesting potential for the energy economy under local geological conditions.

Independently of these positive developments, the experience with established and currently-planned projects shows that there is still a great need for further research and development work. This is especially true of electric power generation using geothermal resources. At our latitudes, that requires accessing of geothermal resources at a depth of 4 to 5 kilometers. With the current state of the technology, this involves risks and high initial investment costs, and demands site-specific solutions to problems. In order to pave the way for industrial-scale utilization of geothermal resources, technologically and economically feasible con-

Fig. 8 *The installation of piping to secure the shaft is part of the so-called borehole completion* (photo: GFZ).

cepts and strategies must be developed which will guarantee secure planning and economically competitive operation of the plants. Only in this way will they offer a solid basis for project investments.

Systematic further research and development approaches with emphasis on reducing costs and risks as well as accessing geothermal reservoirs in a way which will maximize productivity are of prime importance for progress in geothermal energy usage. The research results and experience gained at reference sites such as Gross Schönebeck or the European HDR Pilot Project in Soultz-sous-Forêts, Alsace (France) represent global benchmarks for research and development on EGS systems. Processes and methods which are succesful at these sites can be utilized worldwide at sites with similar geological features. They can contribute to strengthening the as-yet minor contribution of geothermal resources to the world's energy supply.

Summary

The increasing calls for base-load power generation from sustainable energy sources and ambitious goals for climate protection require long-term and effective development of geothermal applications. The complexity of geothermal systems makes a holistic approach necessary, taking into account the interactions of all the components, from accessing the reservoir to the generation of electric power. There is need for research in the technology for locating and engineering the reservoirs in order to increase their productivity, thus achieving cost-effective operation of the plants. Furthermore, the above-ground components must be optimized in terms of performance. Demonstration projects play an especially important role in this process. In Central Europe, the Enhanced Geothermal Systems technologies (EGS) are at the focus of development. They apply hydraulic stimulation to artificially increase the water permeability and heat-exchange area of the deep rock formations and thereby improve the productivity of the reservoirs. EGS technology is particularly suitable for sites which are not cost effective in their natural state. Around 95 % of the reservoirs in Germany could be exploited with this technology.

Acknowledgments

For their support of important subprojects, we thank the German Federal Ministery for the Environment, Nature Conservation and Nuclear Safety (BMU), as well as the Federal Ministry of the Economy and Technology (BMWi).

References

[1] R. Goldstein *et al.*, *Geothermal Energy*, in: IPCC Special Report on Renewable Energy Sources and Climate Change Mitigation (Eds. O. Edenhofer *et al.*), Cambridge University Press, Cambridge, UK and New York, NY USA, **2011**.

[2] R. Wiser *et al.*, *Wind Energy*, in: IPCC Special Report on Renewable Energy Sources and Climate Change Mitigation (Eds. O. Edenhofer *et al.*), Cambridge University Press, Cambridge, UK and New York, NY USA, **2011**.

[3] Geothermal Information System for Germany, www.geotis.de/index.php?loc=en_us.

[4] German Federal Ministry of Commerce and Technology, *Energy Data*, Complete Edition, Table 6. Download from: www.bmwi.de/BMWi/Navigation/Energie/Statistik-und-Prognosen/energiedaten.html (in German; for English site, see: www.bmwi.de/English/Navigation/energy-policy.html).

[5] Report of the German Federal government on a concept for the promotion, development and marketing of geothermal power generation and heat production, May **2009**, p. 10. Download from: www.erneuerbare-energien.de/inhalt/43494/4590 (in German).

[6] S. Hurter and R. Schnellschmidt, Geothermics **2003**, *32* (4–6), 779. Further reading: *Geothermal Energy Systems: Exploration, Development and Utilization* Ed. Ernst Huenges, Wiley-Blackwell **2010**.

About the Author

Ernst Huenges, physicist and process engineer, is leader of the International Centre for Geothermal Research and the section Reservoir Technologies at the Helmholtz Centre Potsdam GFZ – German Research Centre for Geosciences. Currently, he is chairman of the Research program Geothermal Technology of the Helmholtz Association and leader of various geothermal projects.

Address:

Prof. Dr. Ernst Huenges, Helmholtz Centre Potsdam, GFZ – German Research Centre for Geosciences, International Centre for Geothermal Research, Telegrafenberg, 14473 Potsdam, Germany. huenges@gfz-potsdam.de

Biofuels
Green Opportunity or Danger?

BY ROLAND WENGENMAYR

Biofuels of the first generation show a sobering balance: They don't necessarily release a smaller amount of greenhouse gases than fossil fuels. Large-scale cultivation of energy-yield plants in addition threatens food production and ecosystems. Hopes now rest on the second generation – and on completely new processes.

Within a short time, the image of biofuels plummeted from being a beacon of hope to a suspected cause of famine and the destruction of ecosystems. Detailed analyses have revealed just how complex the social and economic consequences of an energy harvesting method that requires great areas of valuable farmland for its large-scale deployment can be. In addition, the conversion of primeval forests, natural CO_2 sinks, into cultivated areas worsens the climate balance [1,2]. Furthermore, intensive cultivation of energy-yield plants, e.g. corn or sugar cane, in regions of low rainfall also burdens the water supply, as has been shown for example by investigations in China and in India. The Indian government has in the meantime abandoned its goal of increasing the proportion of biofuels used from 5 % to 20 % of the total vehicle fuel consumption by 2012, in view of their competition with the cultivation of food plants.

At least today's biofuels of the so-called first generation exhibit a precarious balance. These include bio-alcohol (ethanol) from starchy plants and biodiesel from oil-containing high-yield energy plants (Figure 1). The climate balance of some production systems can even be worse than that of the fossil fuels they were supposed to supplant. A major portion of the blame for this results from the use of massive quantities of nitrogen fertilizers for cultivating the energy-yield plants; the fertilizers release large amounts of the greenhouse gas nitrous oxide ('laughing gas'). The atmospheric chemist Paul Crutzen (winner of the Nobel Prize for chemistry in 1995) calculated that biodiesel from rapeseed alone is up to 1.7 times more harmful in terms of climate change than fossil diesel fuels at their best [3].

Just how critical the situation is can also be seen e.g. from the disaster surrounding the E10 motor fuel in Germany. A major portion of German drivers have boycotted this fuel out of fear that it could damage their engines; but they are confirmed in this behavior by the generally poor image of biofuels. E10 stands for a 10 % admixture of bio-ethanol in the usual gasoline, while E25, which has been used in Brazil since the late 1970s, has 25 % added ethanol.

FIG. 1 | BIOFUELS OF THE FIRST GENERATION

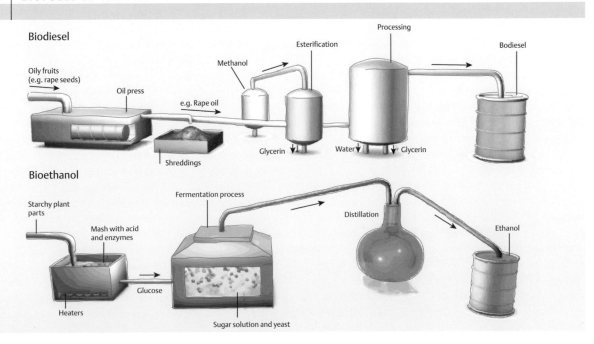

Biodiesel — Oily fruits (e.g. rape seeds) — Oil press — Shreddings — e.g. Rape oil — Methanol — Esterification — Glycerin — Water — Glycerin — Processing — Bodiesel

Bioethanol — Starchy plant parts — Mash with acid and enzymes — Heaters — Glucose — Fermentation process — Sugar solution and yeast — Distillation — Ethanol

Above: The scheme for production of biodiesel (methyl ester) from oil-containing fruit bodies, e.g. rape seeds or soybeans. The conversion plant uses catalysis to produce methyl ester from the plant oils and methanol; the processing plant then purifies it. Below: Production of bio-ethanol from starchy plant parts (sugar beets, potatoes, grain etc.).

Renewable Energy. Edited by R. Wengenmayr, Th. Bührke. Copyright © 2013 WILEY-VCH Verlag GmbH & Co. KGaA, Weinheim

FIG. 2 | BIOMETHANE AND BIOFUELS OF THE SECOND GENERATION

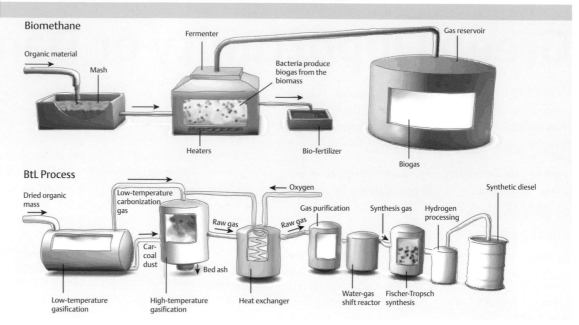

Above: The principle of production of biogas from starchy plants and organic refuse – the spectrum ranges from energy corn to manure. Biogas consists of 50–70 % methane, while the rest is mainly CO₂. The latter is separated out when the biogas is to be used as vehicle fuel (not yet mature for commercialization). Below: Production of synthetic diesel (and related fuels) using the biomass-to-liquid process, starting from dried plant material (straw, wood shavings, etc.).

In the case of diesel, B7 for example stands for a 7 % fraction of biodiesel. In the face of massive criticism from the OECD and from NGOs, the European Union has in the meantime revised its goal of increasing the proportion of biofuels in the overall vehicle fuel consumption to 10 % by 2020. The goal still stands, but only insofar as 40 % of these biofuels must be from the so-called second generation. These are synthetic fuels which can be produced from various kinds of biomass, including plant refuse, manure and household garbage. Figure 2 shows also how bio-methane is produced; it is occasionally included among the biofuels of the second generation.

The energy and climate-change balance of the different generations of biofuels from the sun to the fuel tank will be discussed comprehensively by Gerhard Kreysa in the following chapter. He also makes an estimate based on the current state of knowledge of just what contribution biofuels can make to the future energy supply for humanity. In spite of all the criticisms, they still have the potential to contribute a socially and ecologically acceptable proportion of the energy consumed for transportation. This is especially important for the fast-growing area of air travel, since for aircraft, practically only liquid fuels can be used, owing to their high energy densities.

Two later chapters also deal with the future of biofuels. Nikolaus Dahmen and his co-authors from the Karlsruhe Institute of Technology (KIT) in Karlsruhe, Germany, introduce their patented bioliq® process for the production of second-generation synthetic fuels. This process is already well on the way to being commercialized. In contrast, biofuel from algae is still a dream for the future; it would require no valuable farmland. Carola Griehl and her co-authors from the Anhalt University of Applied Sciences in Köthen, Germany, and the KIT summarize the state of re-

search in this field. For future vehicle fuels from a wide variety of bio-materials, a number of other fascinating ideas and research projects are in the works worldwide. The remainder of this chapter describes some interesting examples.

Ethanol from Cellulose

In the long term, the production of biofuels could be revolutionized by technologies which open up new methods for using resources or make use of completely new resources. One of these possibilities is cellulose. The forests of the earth produce 10^{12} tons of cellulose annually from CO_2 and water during their growth phases [4]. Cellulose could become one of the major energy sources in the future.

Chemically, cellulose is a polysaccharide, a polymer sugar molecule containing many monomer units. It can in principle be split up into its simpler monomer sugars, e.g. glucose. The latter can then be fermented to yield alcohol as a fuel component (pulp alcohol). However, cellulose is chemically very stable. Current technologies for splitting it must overcome its stability. They first convert wood into pulp; it is then split chemically by sulphuric acid. However, the acid destroys a part of the sugar molecules obtained in this process. Furthermore, a major portion of natural cellulose cannot be split in this way, namely monocrystalline cellulose, which is particularly stable [5]. As a result, the current technology is too inefficient for fuel production.

Scientists at the Max Planck Institute for Carbon Research in Mülheim, Germany, have therefore developed an alternative process which is considerably more efficient [5]. First of all, it makes use of the discovery that the liquid salt alkyl methyl imidazole, a so-called ionic liquid, can dissolve cellulose. This dissolved cellulose must then be decomposed into glucose. This is accomplished by an acidic solid-state catalytic system, which was developed in Mülheim. It decomposes the long-chain molecules into short chains and can then be readily recovered from the reaction products. In the final step, enzymes cleave these short chains into simple sugars.

This Mülheim Process is very efficient and can split even monocrystalline cellulose. An additional advantage is that it can be applied to many types of plant remains, even to untreated wood. However, it will require further research before it is ready for use on a broad technical scale. A major problem is the high price of the ionic liquids; in an industrial process, they would therefore have to be recovered to a large extent and recycled.

Genetic Technology for Biofuels

Around the world, a number of research groups are working on the idea of employing genetically-modified microorganisms to produce biofuels. The goal is to convert the bio-material into fuel in the gentlest possible manner with respect to resource consumption, for example via a highly efficient utilization of biogenic wastes.

A prominent example demonstrates the enormous interest in the applications of genetic technology on the part of industry: the Energy Biosciences Institute at the Lawrence Berkeley National Laboratory in California, founded in 2008 by the Nobel Prize winner and current US Minister of Energy Steven Chu, is being co-financed by the petroleum concern BP with a half-billion dollars [6]. There are already numerous smaller start-up firms in this area, and not only in the USA.

An example is the *Butalco* company in Zug, Switzerland, co-founded by the microbiologist Eckhard Boles from the University of Frankfurt. One of its projects involves genetically modified yeasts, which in fermentation yield the more energy-rich butanol instead of ethanol. This should permit an increase in the energy efficiency of the fermentation process by 40 %. The yeast must however first be made resistant to high butanol concentrations. The *Butalco* process in addition chemically cleaves the cellulose in plant waste, for example straw, by the addition of hydrochloric or sulphuric acid, and thereby makes it accessible to fermentation by the modified yeast. *Butalco* has operated a pilot plant since 2010, together with the University of Hohenheim, near Stuttgart [7].

Splitting cellulose into exploitable sugars is thus not only a goal of catalysis researchers, but also of genetic technologists. At the Technical University in Vienna, for example, the research group of the process technologist Christian Peter Kubicek has been experimenting for some time with the mold fungus *Trichoderma reesei*. With the enzymes that it produces, this mold can effectively decompose cellulose-containing fibers into sugar. Kubicek's group has now succeeded in decoding the genome of this mold. They hope to be able to optimize it for fuels production.

Alongside bio-alcohol petrol replacements, the "genetic energy technologists" are naturally also working on diesel or kerosene manufactured by microorganisms. One candidate is genetically modified *coli* bacteria, which can produce biodiesel-like esters. The systems biologist Uwe Sauer from the Federal Polytechnic University in Zurich carries out research on them, for example. He consults for the Californian start-up *LS9*, which is experimenting with small pilot plants [8].

The new field of synthetic biology is pursuing the radical and distant goal of designing completely new microorganisms for fuel production. This will certainly remain in the realm of science fiction for some time to come. Even the current research which uses existing microorganisms is constantly running up against the problem that manipulations of the genes regulating their metabolism causes complex side effects. These can give rise to unfavorable changes in the characteristics of the microorganisms, in the worst case making them non-viable.

These ideas have stirred up discussion as to whether such modified or completely new organisms might have negative effects on the environment if they should somehow be released from the bioreactors. Supporters of the research respond to this criticism by noting that the artificial organisms would hardly be viable in the – for them – harsh natural environment. Nevertheless, it is in the interest of future fuel producers to carefully and critically consider the risks entailed by such radical new production methods. After all, the current situation of the producers of first-generation biofuels offers an example of how quickly public opinion can change direction, followed by that of the politicians.

Summary
Today's first-generation biofuels exhibit a critical balance. Cultivation of energy-yield plants consumes valuable farmland and can cause famines in poorer agricultural countries and destroy irreplaceable biotopes. Furthermore, the massive use of nitrogen-containing fertilizers can contribute to a serious greenhouse effect through emissions of nitrous oxide. Hope for the future is based on second-generation biofuels and biogas, which can be produced from various biogenic wastes. In the more distant future, cellulose or algae could provide the raw material for biofuels. Genetically modified microorganisms could also be employed in biofuels production.

References
[1] J. Fargione et al., Science 2008, 319, 1235.
[2] T. Searchinger et al., Science 2008, 319, 1238.
[3] P.J. Crutzen et al., Atmos. Chem. Phys. Discuss. 2007, 7, 11191. Online: www.atmos-chem-phys-discuss.net/7/11191/2007.
[4] D. Klemm et al., Angew. Chemie Int. Ed. 2005, 44, 3358.
[5] R. Rinaldi, R. Palkovits, F. Schüth, Angew. Chemie Int. Ed. 2008, 47, 9092.
[6] berkeley.edu/news/media/releases/2007/02/01_ebi.shtml.
[7] www.chemicals-technology.com/projects/butalcocellulosiceth.
[8] N. Savage, Technology Review 2007, 6. Online: www.technologyreview.com/Biztech/18827.

Biofuels are Not Necessarily Sustainable

Twists and Turns around Biofuels

BY GERHARD KREYSA

Germany is the European champion in the production of biodiesel, by a large margin [1]. The fraction of the total European production (EU-27) of 5.71 million liters in 2007 that was manufactured in Germany amounted to 50.7 %, a factor of 3.3 more than France, second on the list. Of the 6.29 million tons of rapeseed harvested in Germany in 2009, 38 % came from the three states of Mecklenburg-Western Pomerania, Saxony-Anhalt, and Brandenburg [2]. Thus, the land of the famous German novelist Theodor Fontane has become a world center of biodiesel production. But this development, which began with great euphoria and sounded like a German success story of sustainable resource usage, now bears a number of ecological and economic question marks and has become a sensitive topic of political discussions. A peculiar, almost tragic aspect of this saga is the fact that thoroughly reasonable and honorable technical, ecological and economic motives led in the end to a situation in which the resulting reality largely undermines the original goals. We consider in this chapter how this could happen and discuss the case in some detail.

Two driving forces have propelled the idea of biofuels: On the one hand, the limited future for the exploitation of fossil fuels, which today still dominate our energy supply; and on the other, the steady rise of the CO_2 content of the earth's atmosphere which is caused by burning those fuels, and its role in climate change via the greenhouse effect.

The Rational Basis – the Carbon Cycle

We dispense here with the usual discourse flaunting the latest numerical data on the amounts of coal, petroleum and natural gas remaining underground, the difference between reserves and resources, and the relation between the amounts already consumed and those remaining (peak-oil discussion); instead we refer the reader to the literature [3]. The danger is still too great that such numbers would have to be updated, as were the estimates of the Club of Rome [4], which proved to be too low. The secure reserves of all fossil primary products today, in spite of intensive extraction in recent decades, are still two to four times greater than at the time of the first oil crisis in 1973 [5]. This at first surprising statement becomes understandable if one remembers that the expensive business of exploration for new deposits is financed by only a rather small portion of the profits from fossil raw materials production. In a market-oriented system, no generation is willing to accept high price supplements for raw materials to finance exploration for the benefit of arbitrarily many future generations. These system-immanent uncertainties however in no way modify the fact that all of our raw materials are available only in finite quantities.

FIG. 1 | THE RAPID CARBON CYCLE ...

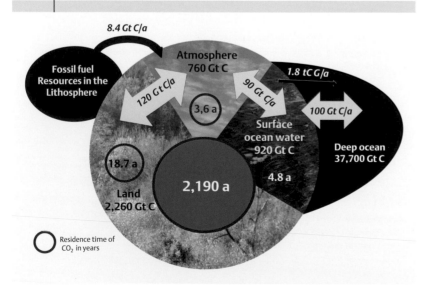

8.4 Gt C/a

Fossil fuel Resources in the Lithosphere

Atmosphere 760 Gt C

120 Gt C/a

90 Gt C/a

1.8 tC G/a

3,6 a

Surface ocean water 920 Gt C

100 Gt C/a

Deep ocean 37,700 Gt C

18.7 a

2,190 a

4.8 a

Land 2,260 Gt C

Residence time of CO_2 in years

... with its linked long-term reservoirs. The figure shows the carbon content of the reservoirs, reversible flows and uni-directional flows (italics), and the derived residence times for CO_2 in each compartment and in the whole cycle (a: year).

If we leave the superficially-correct economic deliberations out of consideration for the moment, then it would actually be preferable if we were facing immediate exhaustion of our fossil raw materials. The greater the *de facto* reserves, the more irrational is their complete exploitation, since their combustion will shift the earth's atmosphere back to a state which it once had, long before the appearance of human beings, and for which there is not the least evidence that it would be compatible with our continued existence. Figure 1 shows the carbon cycle in a somewhat unusual depiction, which is limited to those components that together can be termed the 'fast carbon cycle'. The numbers there for the carbon content of the reservoirs and its rate of flow between them were adopted from the IPCC [6]. The net flows between the components *atmosphere*, *land*, and *surface oceans* (down to a depth of ca. 700 meters) are in the range of 1 Gt of C/a and demonstrate that equilibrium has not been established between these reservoirs. The most important perturbing factor is the anthropogenic input of around 8 Gt C/a into the overall cycle due to the combustion of fossil raw materials.

In contrast to the unidirectional net flows, there is a very rapid, bidirectional CO_2 exchange between the atmosphere and the land as well as the surface oceans. Considerations similar to those applied to reaction kinetics indicate that the residence times in these coupled reservoirs are quite short and are in the range of 3.6 to 18.7 years.

It is much more enlightening, however, to look at the overall cycle instead of considering the individual reservoirs. For the residence time of carbon in this cycle as a whole, the net drain of 1.8 Gt C into the deep oceans is relevant, and leads to a value of 2190 years. The significance of this statement becomes apparent through a comparison with nuclear power. For the direct final storage of spent nuclear fuel elements, the lifetime of the radiotoxicity is over 100,000 years. If, however, the fuel elements are re-processed and their long-lived plutonium is recycled back to the power reactors, this lifetime is reduced to around 1,000 years [7]. This comparison shows that waste disposal of anthropogenic CO_2 confronts humanity with similar problems to those from the use of nuclear power. Wolf Häfele, a former chairman of the board of the Jülich Research Center, often spoke even 20 years ago of the problem of 'fossil waste disposal', to point up this analogy. In recent years, several model calculations have been published [8-10], and they all confirm the extremely slow natural decay constant for CO_2 which has been introduced into the cycle.

The anthropogenic fossil input amounts to only 7 % of the exchange flow between the atmosphere and the land. The value 120 Gt C/a corresponds to the annual photosynthesis activity through assimilation. The equally large backflow is due to roughly equal parts of respiration and microbial decomposition [11]. The fact that photosynthesis exceeds the anthropogenic CO_2 production 19-fold has for decades been a strong motivation for substituting fossil energy carriers by raw materials from the biomass, since the latter are produced by growing plants. Liquid biofuels seem especially attractive, offering a similarly high energy density to that of the fossil fuels, and requiring no essential modifications of the infrastructure for motor transportation. Brazil took on the role of trailblazer in this area more than 20 years ago with its use of bioethanol. In 2007, the fraction of biofuels used in Germany as primary motor fuels already amounted to 7.3 %, of which three-fourths consisted of biodiesel [1].

In Table 1, the various types of biofuels, their chemical constituents, their raw materials, conversion processes and their development status are shown comparatively. Only the fuels of the first generation are at present technically available on a large, industrial scale. Worldwide, bioethanol and biodiesel are quantitatively predominant. Now that the

TAB. 1 | PRODUCTS AND PROCESSES FOR BIOFUELS

Biofuel	Chemical Constituents	Raw Materials	Process	Development Status
First Generation				
Biodiesel	Fatty-acid methyl esters	Fats and oils	Esterification	Commercial
Bioalcohols	Ethanol, butanol	Starch, sugars	Fermentation	Commercial
ETBE	Ethyl-t-butyl ether	Starch, sugars	Fermentation/synthesis	Commercial
Second Generation				
BtL	Hydrocarbons	Lignocellulose	Gasification, Fischer-Tropsch synthesis	Pilot plants
Methanol	Methanol	Lignocellulose	Gasification, synthesis	Pilot plants/research
DME	Dimethyl ether	Lignocellulose	Gasification/methanol/synthesis	Pilot plants/research
Bioalcohols	Ethanol, butanol	Cellulose, hemicellulose	Digestion/fermentation	Pilot plants/research
Tailor-made Fuels and Fuel Constituents				
Synfuel	Hydrocarbons	Lignocellulose	Gasification/methanol/synthesis	Pilot plants/research
	Lactones, ethers, furanes etc.	Cellulose, hemicellulose	Synthesis	Basic research
	Aromatics	Lignin	Synthesis	Basic research

The features of various concepts for manufacturing liquid fuels from the biomass, after Leitner [12].

initial euphoria has evaporated, the experts are largely in agreement [12-16] that the biofuels of the first generation can at best be seen as an interim solution, particularly in Europe, and for a variety of reasons cannot be considered sustainable.

The destruction of rain forests to obtain land for cultivating oil plants, which are for the most part then used to produce biofuels, is particularly absurd and reprehensible. This fuel is even exported to Europe! Here, again, a detailed examination of the carbon cycle makes the danger apparent: of the roughly 2,260 Gt of carbon stored on the land surface of the earth, only about one-fourth is in the plants; the remainder is mainly stored in humus near ground level [11]. When the forest is cleared, this humus is for the most part converted rapidly and irreversibly back to CO_2. Since the later biomass-biofuel cycle is at best neutral with respect to CO_2, the net result is an extensive shift of carbon from the land into the atmosphere and the oceans.

Hopes are thus currently focused on the second-generation biofuels. The most technically advanced is the total gasification of biomass and subsequent synthesis of liquid fuels using the Fischer-Tropsch process. Recently, clear-cut progress has also been made in the fermentative production of bio-alcohols from cellulose and hemicellulose. The digestive pulping of lignocellulose is proving to be more difficult, and it is the subject of intensive research.

A Critical Discussion of Several Criteria

Fundamentally, every potential technical solution for the substitution of fossil-fuel combustion must meet several criteria, which determine whether it is reasonable to carry it through. In the following, for clarity we distinguish between elimination criteria and economic criteria. A first elimination criterion is the carbon balance with respect to the greenhouse effect. It must include the entire process chain and all of the relevant greenhouse gases (GHG). Repeatedly, GHG balances have been published, even by reputable organizations such as the OECD, which left certain aspects out of consideration [17]. We have Crutzen [18] to thank for first compiling GHG balances for biofuels and comparing them; in particular, for taking into account the emissions due to changes in land use as described above (which were not included e.g. in [1]), and also the release of N_2O from the necessary fertilizers. His results are summarized in Figure 2.

The **relative greenhouse effect** gives a quantitative description of the greenhouse effect which would result from utilization of a biofuel, including the whole production chain from cultivation of the plants to combustion of the fuel, relative to that resulting from the same energy released by combustion of a fossil fuel (diesel). The greenhouse effect refers to the storage of solar heat in the earth's atmosphere. Since the release of N_2O, whose greenhouse effect is nearly 300 times stronger than that of CO_2, is subject to considerable fluctuations depending on weather conditions and fertilizer usage, the results of such a comparison have wide margins of uncertainty. Nevertheless, the comparison shows that with the exception of bioethanol from sugar cane in Brazil, the greenhouse effect from biodiesel and bioethanol is greater than that from the CO_2 released by burning conventional fossil fuels. This is not acceptable in the long term and represents an effective elimination criterion, assuming that further careful investigations, which are extremely complex, confirm these results. The probability of this is high.

An elimination criterion which is somewhat more readily accessible is the **net energy gain**. This quantity requires some explanation, since, according to the First Law of Thermo-dynamics, it should always be negative, because energy cannot be created, only converted to other forms, and that is always accompanied by losses. Even the concept of 'energy loss' is thermodynamically not admissible; generally, it refers to 'waste heat' which cannot be used in the process considered and results as an unwanted byproduct from energy conversions.

In considering sustainable energy sources for the substitution of fossil energy carriers, it is therefore usual not to take into account the solar energy used, either directly or indirectly (e.g. as the potential energy of water utilized for hydroelectric power generation). For biofuels, the area-based net energy gain results from the harvest yield of biomass per hectare and the corresponding fuel equivalent (i.e. the amount of diesel fuel which would release the same heat of combustion as the biofuel manufactured from that biomass), minus the overall input of energy from fossil fuels (from the fertilizer factory through land preparation and cultivation on to the biofuel tank) required for the manufacture and use of the biofuel. The energy gain thus defined

FIG. 2 | THE RELATIVE GREENHOUSE EFFECT OF BIOFUELS ...

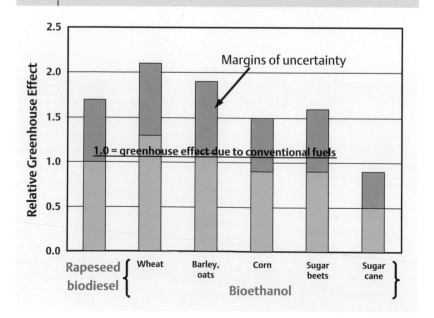

... including the release of nitrous oxide from nitrogenous fertilizers (data from [18]).

can indeed be negative, and in that case it represents a clear-cut elimination criterion.

To be sure, up to now there have been relatively little reliable data on this quantity published in the literature [19], and careful calculations of the energy balance for the many variants of cultivation and manufacture of biofuels are still lacking. Therefore, the Agency for Renewable Resources (FNR) [1] gives no data on the net energy gain. Instead, it makes use of a somewhat different definition of the net energy yield. The gross energy yield denotes the energy content of the biomass harvested per hectare and year. The (gross) fuel yield [1] represents the energy content of that amount of fuel which can be manufactured from the harvested biomass. The **net energy yield** [1], on the other hand, is a measure of the energy from non-renewable sources which can be saved by using the biofuel instead. It corresponds to the energy content of the fossil fuel saved (i.e. gasoline or diesel fuel plus the energy required for its production and distribution), minus the total of the fossil energy consumed for the overall production of the biofuel. The energy content of coupled production such as coarse colza meal or glycerin is taken into account and contributes to a positive balance. With a view to the potential for substituting fossil fuels and reducing carbon dioxide emissions, this quantity makes sense. However, as explained above, it is not sufficient as a criterion for evaluating the greenhouse effect.

The net energy yields published by the FNR [1,20] can be used to make a comparison between varous methods of biofuels manufacture and alternative energy-related concepts for use of the land. Figure 3 shows the results of such a comparison. There, we have computed how far one could drive an automobile using the net energy yield from one hectare of agricultural land (a diesel consumption of 5 l/100 km, corresponding to 490 Wh/km, was assumed for the automobile). This cruising range provides a meaningful efficiency criterion.

As one can see from the figure, the biofuels of the first generation (bioethanol and biodiesel) give a range of around 23,000 km. From the biofuels of the second generation, which are for the most part still under development [12], about 65,000 km can be expected. An only slightly lower range can already be obtained with technology available today for biomethane production.

However, it is particularly interesting to consider another alternative. Instead of growing energy-yield plants, one cultivates rapid-growth plants (such as poplar or willow trees) or miscanthus (miscanthus sinensis). On the average, an amount of biomass can be produced in this way with a heat content of 55 MWh per hectare (compare the data in [21]). The resulting biomass can be mixed without problems into the coal burned by a modern coal-fired power plant (efficiency 45 %), generating electric energy amounting to 24.8 MWh$_e$ in a CO_2-neutral manner. The resulting cruising range of 123,800 km is found by assuming 80 % efficiency for storage batteries and an energy consumption of

160 Wh/km for the electric automobile. Analogously to BtL (biomass to liquid), this method of using the biomass as fuel is termed BtE (biomass to electricity). Independently of the three computations of the present author [22] shown in Fig. 3, the FNR has published a similar figure [20] (without the BtE variant), where the values differ only slightly and which meanwhile can be seen in Wikipedia [23].

Based on the data from the most recent FNR study [1], Table 2 sets out and compares the various possibilities for converting solar energy into traction. As emphasized above, the cruising range as computed from the net fuel yield is a meaningful efficiency criterion. This range is also directly proportional to the reduction in CO_2 emissions achieved. Regarding avoidance of CO_2 emissions, biomass to electricity on the basis of miscanthus is thus 6.4 times more effective than the production of biodiesel from rapeseed. If one uses palm oil instead, the effciency is doubled. This is the origin of the rather disastrous developments mentioned above: Many developed countries began importing palm oil from developing countries with the goal of efficient production of biodiesel. Palm oil in the meantime dominates worldwide production of plant oils, with an annual harvest of 45 million tons, of which 87 % originates in Indonesia and Malaysia [2]. In Germany, two-thirds of the rapeseed oil from the annual harvest of 6.3 million tons of rapeseed is converted to biodiesel. In 2009, the EU-27 countries consumed 16 million tons of oil seeds for bioenergy; this is nearly equal to the whole amount imported, 17.3 million tons. In order to free up land for palm-oil cultivation, enormous areas of tropical forest were cleared. This change in land usage released large amounts of carbon dioxide which was previously bound in the biomass and especially in humus. It has been pointed out by Michel [25] that it will take 425 years for CO_2 fixation processes to bind the CO_2 that was set free by clearing rainforest land for palm-oil cultivation.

FIG. 3 | CRUISING RANGES OF AUTOMOBILES ...

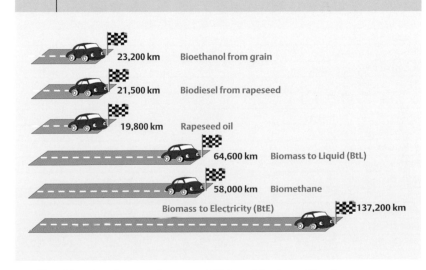

... achievable using the annual energy yield from 1 hectare of land.

These problems have been recognized by a number of countries. National governments and the EU are therefore working together to give tax advantages only to biodiesel which was produced in a sustainable manner.

The direct utilization of solar energy via photovoltaics or solar-thermal plants is considerably more efficient than any method based on photosynthesis can be. The poor efficiency of photosynthesis, around 0.5 % (next-to-last column in Table 2), which in effect should obviate the use of the biomass as an energy carrier, especially in the form of first-generation biofuels, was emphasized by Michel [25]. He also emphatically dampened any hopes of being able to increase the efficiency of photosynthesis by genetic manipulation. For comparison to the biofuels, therefore, the cruising range of an electric auto powered by direct usage of solar energy via photovoltaic cells is also given in Table 2. In this estimate, the data refer to an existing solar park in Lieberose [24] with an efficiency of only 10 %, and it was also taken into account that only about 0.3 ha of module area can be arrayed on one hectare of land.

The last column in Table 2 shows the overall efficiency for the use of solar energy to power road vehicles, taking into account the efficiencies of the different drive trains

LARGE AREAS OF RAIN FOREST ARE BEING CLEARED TO MAKE LAND AVAILABLE FOR PALM-OIL CULTIVATION

(gasoline engine, diesel engine, electric motor). The clear superiority of photovoltaics is apparent. It becomes notably higher when one takes account of e.g. the module area on rooftop installations and considers the more expensive, but already available solar cells with still higher efficiencies.

Finally, it is of interest also to ask the question as to whether the available potential for the use of bioenergy is sufficient to cover the energy demands of humanity. In an earlier publication [5], the author has already shown that the bioenergy available from the entire area of land currently in use for agriculture on the earth (measured as the energy content of the biomass produced annually) just corresponds approximately to the current annual worldwide energy consumption. This makes it clear that it is impossible to supply the energy needs of the human race completely from bioenergy.

Considerations of this kind have given rise to a clear-cut rethinking, at least among political advisors. The WGBU (Scientific Council on Global Environmental Change of the German Federal government) recommended at the end of 2008 that subsidies of liquid fuels be terminated and the money spent instead on electric vehicle research and development [26]. We quote here some of the reasons given in the text:

- "Technological policies on the usage of bioenergy for transportation must be redirected. Subsidies of liquid biofuels for motor vehicles cannot be justified in terms of sustainability criteria, particularly in industrialized countries."
- "The quotas for mixing of biofuels into gasoline and diesel fuels should not be increased, and the current addition quotas should be reduced to zero over the next three to four years."
- "The highest efficiencies in the use of the biomass as an energy source for transportation are obtained by electric power generation and the use of electrically-powered vehicles. Suitable guidelines for the support and expansion of electric transportation should be put in place."

The urgency of following these recommendations is reinforced by the concerns which are increasingly expressed not only in the public media, but also in the professional literature [15], arising out of a critical consideration of the competition between the cultivation of energy-yield plants and food production, and in light of the continuing growth of the human population. The suggestions of the WBGU regarding increased subsidies for electric transportation have meanwhile been taken up by the German Federal government, not only in its support for research, but also in the framework of the Economic Stimulus Package II. The political will to opt out of support for liquid biofuels is, in contrast, visible at most in small gestures. A weak but still noticeable signal went out from the Energy Taxation Act,

TAB. 2 | COMPARATIVE CONSIDERATIONS

Solar-Energy Usage	Gross Energy Yield [21] (MWh$_{th}$/ha)	Fuel Yield (MWh$_{th}$/ha a)	Net Fuel Yield (MWh$_{th}$/ha a)	Cruising Range (km/ha a)	Photosynthesis efficiency (%)	Sun – Wheel efficiency (%)
Bioethanol from grains	51.00	15.28	14.44	23185	0.510	0.038
Bioethanol from sugar beets		36.67	33.33	53505		0.087
Bioethanol from sugar cane		37.50	32.22	51721		0.084
Bioethanol from corn		21.94	11.11	17835		0.029
Bioethanol from lignocellulose		5.83	5.00	8026		0.013
Biodiesel from rapeseed oil	18.30	14.44	10.56	21542	0.183	0.034
Biodiesel from palm oil		40.00	20.83	42517		0.067
Biodiesel from soybean oil		5.83	5.56	11338		0.018
Rapeseed oil		14.72	9.72	19841		0.031
BtL (Dena. Choren)		37.50	31.67	64626		0.101
Biomethane		49.44	36.11	57963		0.094
Biohydrogen		45.00	33.33	53505		0.087
Fast-growth wood with BtE	51.00	23$_{el}$		114750	0.510	0.184
Miscanthus with BtE	61.00	27.5$_{el}$		137250	0.610	0.220
Photovoltaics (data from [24])	10.000	321$_{el}$		1.56 Mio		2.568

(Data from [1] unless otherwise indicated.)

which includes a stepwise increase in energy taxes on biodiesel and plant oils over the coming years [1].

As yet, the errors with respect to biofuels have not been universally perceived and above all have not been consistently corrected. Some scientists and in particular some politicians are again raising hopes whose seriousness appears questionable at best. The newest rescue scheme for bioenergy involves the use of algae. However, well-grounded studies from industry [27] indicate that optimism is misplaced here, since even very favorable assumptions lead to a negative net energy gain, i.e. the production of biofuels from algae requires more energy – without taking into account the solar energy input – than is stored in these fuels. An additional and very plausible confirmation of this statement results from considerations undertaken by Buchholz [28]. If one assumes that algae can be grown up to a dry weight of 10 g per liter of water, and that 70 % of this dry material consists of energetically utilizable hydrocarbons, then one is dealing with a 0.7-percent solution. Oil sands which can be obtained by strip mining are currently considered unprofitable if they contain less than 6 % hydrocarbons. This is by no means an argument against continuing research into algae, with the intensified goal of obtaining valuable and useful natural products; but their use as a source of bioenergy will remain an illusion for a long time to come, if not forever.

Conclusions

To avoid misunderstandings, we note that the considerations on the utilization of the biomass as an energy source and on the production of biofuels given here do not justify a total, unconditional exit scenario from biofuel research and production. Owing to the increasingly intense competition between food production and energy-yield plants for the scarce resources of agriculturally usable land in the future, in the mid- and longer term there will however be no reasonable justification and indeed no necessity to use arable lands for producing biofuels.

In contrast, it is eminently sensible to continue to examine carefully the potential utilization of processes which have already been developed, and those which are still to be developed, for obtaining energy from biogenic wastes of every kind. Of course, here again the requirement that the net energy balance should be positive represents an exclusion criterion. The relevant energy balances which will permit a definite judgment on this question must be determined more extensively and more precisely than has thus far been done.

Furthermore, it has to be remembered that liquid biofuels, like other liquid fuels, offer the advantage of a very high energy density. For a number of applications, this advantage will continue to be essential in the future. An example is air transportation, since it will hardly be feasible

to operate aircraft using batteries or fuel cells. Long-haul freight transport also stretches the limits of electrical mobility, at least until cost-effective fuel cells, including the necessary infrastructure for their fuel supplies, are available. During this interim period, electric vehicles will probably mostly be equipped with a 'range extender' to increase their cruising range; i.e. an internal-combustion engine that operates at constant rpm and generates on-board power for the drive motor. For this purpose, liquid biofuels of the second generation could prove useful. Prognoses in this area are however very unreliable, since all the approaches assume that the mobility behavior of the public will remain unchanged, and this is by no means certain. For example, the electric vehicle of the future could be without exception a rental car, which can be exchanged for another vehicle with a fully-charged battery after its short cruising range of ca. 200 km is spent.

After years of intensive support for research into liquid biofuel production, the lobbyist interests in industry and agriculture have become almost insuperably powerful. This is understandable, particularly from the viewpoint of those who accepted the offers of long-term subsidies in good faith and have on that basis invested in industrial installations. Here, a nearly insoluble conflict has arisen in the short term between the justified claims on legal certainty, and the necessity of ecologically and technically reasonable actions. A one-sided attribution of blame to politicians would however be neither appropriate nor justified. We must not forget that the setting of priorities for the subvention of biofuels, which today seems in retrospect to have been wrong, was not a single-handed political decision, but rather the result of an intensive process of debate and deliberation. Many of the hopes for subvention and support have been fulfilled, but a certain tragedy remains in the fact that ecology, climate protection and social justice were not served well. All those concerned can now only ask that a lesson be learned from such mistakes and that in the future, similar undesirable developments should be avoided by using foresight in making careful evaluations and estimates of the consequences of large-scale actions.

Summary

The annual consumption of CO_2 by photosynthesis is much greater than its anthropogenic production. This suggests that the biomass should be utilized as an energy carrier, and has provided strong motivation for the development of biofuels. In many cases, however, the use of biofuels violates the principles of sustainability. Meeting the world's energy requirements by using the biomass would require all of the available arable land area on the earth. This realization has unleashed a critical discussion of the alternatives 'food or fuel'. From the viewpoint of energy efficiency and ecological compatibility,

DUE TO THE RELEASE OF N_2O FROM FERTILIZERS, A REDUCTION OF THE GREENHOUSE EFFECT OFTEN REMAINS ILLUSORY

the cultivation of rapid-growth wood and its combustion in electric power plants would be preferable to the direct use of biofuels. By far the most efficient utilization of solar power is through photovoltaic cells and solar-thermal plants. However, because of their high energy densities, liquid biofuels of the second generation will find application in the future where electric-powered transport runs up against its limits.

This chapter was printed with the kind permission of *Chemie in unserer Zeit*.

References

[1] N. Schmitz, J. Henke, and G. Klepper, *Biokraftstoffe – eine vergleichende Analyse*, Agency for Renewable Resources (FNR), Gülzow **2009**.

[2] U. Hemmerling, P. Pascher, S. Naß and M. Lau, *Situationsbericht 2010*, Deutscher Bauernverband, Berlin **2009**.

[3] *Energierohstoffe 2009*, Bundesanstalt für Geowissenschaften und Rohstoffe (BGR), Hanover **2009**; and P. Gerling: *Wie lange gibt es noch Erdöl und Erdgas*, Chemie in unserer Zeit, **2005**, *39*, 236.

[4] D. Meadows, J. Randers, and D. Meadows, *Limits to Growth: the 30-Year Update*, Chelsea Green Publishing Co., White River Junction, VT/USA, **2004**.

[5] G. Kreysa, Chem. Ing. Tech. **2008**, *80(7)*, 901.

[6] S. Solomon, D. Qin, M. Manning *et al.*, *Climate Change 2007 – the Physical Science Basis*, Cambridge University Press, NY **2007**, p. 515.

[7] W. Stoll, H. Schmieder, A.T. Jakubick, G. Roth, S. Weisenburger, B. Kienzler, K. Gompper, and T. Fanghänel, *Nuklearer Brennstoff Kreislauf*, Chemische Technik, Vol. *6b*, Metals, Wiley-VCH, Weinheim **2006**, p. 539.

[8] [6], p. 826.

[9] R. Monastersky, Nature **2009**, *458*, 1091.

[10] M.R. Allen *et al.*, Nature **2009**, *458*, 1163.

[11] W. Fritsche, *Auswirkungen der globalen Umweltveränderungen auf die Wertschöpfungen der Natur*, Meeting Reports of the Saxon Academy of Sciences, Leipzig, Mathematical-Scientific Section, **2009**, *131*, Issue 3.

[12] W. Leitner, Kraftstoffe aus Biomasse, Stand der Technik, Trends and Visions, in *Die Zukunft der Energie*, C.H. Beck Publishers, Munich **2008**, p. 190.

[13] G. Sell, J. Puls, and R. Ulber, Chemie unserer Zeit **2007**, *41*, 108.

[14] R. Harrer, Chemie unserer Zeit, **2009**, *43*, 199.

[15] S. Nordhoff, Biotech. J. **2007**, *2*, 1451 and other contributions to this issue.

[16] T. Bley (Ed.), *Biotechnologische Energieumwandlung*, acatech Deutsche Akademie der Technikwissenschaften, Springer Verlag, Berlin, Heidelberg **2009**.

[17] Facts andFigures, Biotech. J. **2007**, *2*, 1460.

[18] P.J. Crutzen, A.R. Mosier, K.A. Smith, and W. Winiwarter, Atmos. Chem. Phys. Discuss. **2007**, *7*, 11191; *ibid.*, **2008**, *8*, 389. Online: www.atmos-chem-phys-discuss.net/7/11191/2007.

[19] B. Schumacher, *Untersuchungen zur Aufbereitung und Umwandlung von Energiepflanzen in Biogas und Bioethanol*, Dissertation, University of Hohenheim, **2008**.

[20] FNR flyer, *Biokraftstoffe Basisdaten Deutschland*, Agency for Renewable Resources, Gülzow, **2007** and **2009**.

[21] FNR flyer, *Biokraftstoffe Basisdaten Deutschland*, Agency for Renewable Resources, Gülzow, **2009**.

[22] G. Kreysa, Contribution to discussion, acatech Workshop of the Topical Network on Biotechnology, 22nd October, **2008**, Berlin.

[23] See under 'Biodiesel' in Wikipedia: de.wikipedia.org/wiki/Biodiesel (In English: en.wikipedia.org/wiki/Biodiesel).

[24] Solarpark Lieberose in Zahlen, **2009**; see: www.solarpark-lieberose.de/zahlen/default.html.

[25] H. Michel, Die Natürliche Photosynthese, Ihre Effizeinz und die Konsequenzen, in *Die Zukunft der Energie*, C.H. Beck, Munich **2008**, p. 71.

[26] *Zukünftsfähige Bioenergie und nachhaltige Landnutzung*, WGBU: Scientific Council on Global Environmental Changes, German Federal government, Berlin **2008**.

[27] U. Steiner, *Betrachtungen zur Energiebilanz und Wirtschaftlichkeit von Biokraftstoffen aus Algen*, 18th–19th Jan. **2010**, Frankfurt am Main.

[28] R. Buchholz, *Perspektiven der Algenbiotechnologie*, BIOspektrum, **2009**, *15*, 802.

About the Author

Gerhard Kreysa was born in Dresden in 1945. Following his studies of chemistry leading to the doctorate, from 1973 on he carried out research at the Karl-Winnacker-Institut of the DECHEMA in Frankfurt/M on problems of technical electrochemistry and environmental protection. He has received a number of scientific awards in Germany and abroad, and was professor at the Universities of Dortmund and Regensburg. From 1992 to 2009, he was Chief Executive of the DECHEMA, and during the same time period served as General Secretary of the European Federation for Chemical Engineering. He is a member of the Swedish Royal Academy of Engineering Science (IVA) and the German Academy of Technical Sciences (acatech). He is concerned with questions of research support, climate protection, and public acceptance of science and technology, serving also in many honorary posts.

Address:
Prof. Gerhard Kreysa, c/o DECHEMA e.V., Theodor-Heuß-Allee 25, 60486 Frankfurt am Main, Germany.
g.kreysa@t-online.de

Biofuels from Algae

Concentrated Green Energy

BY CAROLA GRIEHL | SIMONE BIELER | CLEMENS POSTEN

Out of roughly a half-million species of algae, thus far about 220 macroalgae and 15 microalgae are being used for commercial purposes, for nutrition and animal feed and as ingredients for cosmetics. Given the threatening climate change and increasing scarcity of fossil fuel resources, research is continuing on using algae as alternative energy carriers.

There are an estimated 500,000 different species of algae on the earth. Only a small fraction of these, ca. 220 macroalgae and 15 microalgae, are being utilized commercially for foodstuffs, animal feed, and cosmetic ingredients. Algae are highly interesting as sources of biofuels. First-generation biofuels have come in for criticism, and second-generation biofuels (see the previous chapter), although they can offer a more favorable economic and ecological balance, are limited in their overall production by the available areas of arable land.

Algae, on the contrary, represent no competition to food production, since they can be grown on land areas which are not agriculturally productive. They offer many advantages as compared to terrestrial energy-yield plants; in particular, they produce biomass much more efficiently, with yields of up to 150 tons per hectare annually. In addition, they utilize CO_2 from the air more efficiently than terrestrial plants on a per-hectare basis, and permit the use of CO_2-containing exhaust gases. Closed algae production plants would use less water than conventional agriculture. Effluents or fermentation wastes are good sources of phosphates and nitrates for algae. Not least, the entire biomass of the algae can be utilized, and it can be harvested year-round.

Current research is aimed at making this biomass of the "third generation" ecologically and economically suitable for large-scale practical use. An important criterion for efficiency is the net energy gain, which must be positive (see previous chapter).

A Renewable Resource

Algae are water plants with a simple structure. They occur either as single-celled microalgae which can form colonies, or as many-celled macroalgae, up to 60 meters long (seaweed, kelp). While macroalgae grow mostly in the coastal regions of the oceans, the microalgae, which can take many different forms, usually float among the phytoplankton in ponds, lakes, rivers and oceans.

Like all green plants, algae fill their energy requirements by means of photosynthesis, which releases oxygen. In this process, they utilize the energy of sunlight, and CO_2 as a carbon source to assemble their cellular material (biomass). Photosynthesis produced the oxygen-rich atmosphere of our planet. It is estimated that the plankton algae in the oceans bind $2 \cdot 10^{10}$ tons of carbon annually into their photosynthesis products. They thus take up nearly half of all the CO_2 which enters the atmosphere each year. Without them, the CO_2 concentration in the atmosphere would increase from its current value of 380 ppm to an estimated 565 ppm. For each kilogram of biomass that they produce, the algae consume ca. 1.8 kg of CO_2. This value increases still further if they are induced to make a higher proportion of chemical storage substances such as oils or carbohydrates.

Algae are the most important primary biomass producers on earth. Plankton algae are especially productive: In water with a high concentration of nutrients (nitrates, phosphates), the plankton biomass formed each year is 2-6 tons per hectare, and in the case of an algal bloom, even as much as 60 tons [1]. Likewise, the productivity of the macroalgae is also considerable, amounting to 10-36 tons per hectare and year, and contributes 10 % of the overall marine primary biomass production [2].

Algae thus represent an important potential sustainable resource, which can be enhanced still further by cultivation. Although the efficiency of photosynthesis and biomass production is limited by several factors, in practice currently 2 to 5 % PCE can be attained, depending on the cultivation system. (PCE stands for photo-conversion efficiency and denotes the fraction of the light energy which is converted to chemical energy.) Algae thus utilize the sunlight for the formation of biomass and natural products much more efficiently than terrestrial plants do (PCE < 1 %). For example, with microalgae in systems using open basins, ca. 10–30 (max. 50) tons of dry biomass per hectare and year ($t\ ha^{-1}\ a^{-1}$) can be obtained; in glass tubular reactors, as much as 80-150 (max. 200) $t\ ha^{-1}\ a^{-1}$. The yields from macroalgae grown by aquaculture, ca. 50-85 $t\ ha^{-1}\ a^{-1}$ of dry biomass [3], are notably higher than those from high-yield terrestrial plants (rape: 4.2 $t\ ha^{-1}\ a^{-1}$; miscanthus

Renewable Energy. Edited by R. Wengenmayr, Th. Bührke. Copyright © 2013 WILEY-VCH Verlag GmbH & Co. KGaA, Weinheim

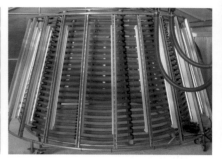

Fig.1 *Production systems for algae, from left: Dunaliella in open ponds in Eilat (Israel); Chlorella in glass tubular PBR in Klötze (Saxony-Anhalt, Germany); Scenedesmus in flexible tubular PBR in Köthen (Saxony-Anhalt, Germany)* (photos: R. Wiffels, Wageningen; Anhalt University of Applied Sciences).

sinensis: 25 t ha^{-1} a^{-1}), which in addition require irrigation and fertilizers.

Production Methods

Microalgae can be cultivated either in basins (open ponds) or in closed systems, i.e. in photobioreactors (PBR, see Fig. 1). Their production costs, depending on the location, the species of algae, and the cultivation system, are currently between 3.5 and 50 €/kg. Due to their low investment and operating costs, at present open pond systems predominate worldwide. In such systems, however, controlled and contamination-free growth is not possible, and the PCE values and biomass concentrations achieved remain modest.

Reproducible, contamination-free and highly productive cultivation demands the use of closed photobioreactors (PCE values up to 5 %). These also require less land area with higher yields. They in addition make more efficient use of CO_2 per volume than the open systems. However, their investment and operating costs are notably higher. There are

tubular and plate reactors, each of which has its advantages and disadvantages. Worldwide, glass tubular systems are preferred for high-quality products. These are utilized in Germany, Israel and India to manufacture algal biomass or the pigment astaxanthin. The world's largest installation is in Klötze, in Saxony-Anhalt, Germany. Its 500 km of tubing contain 600,000 liters and yield ca. 60 tons of biomass in a half-year of operation between spring and autumn.

The Dresden firm *GICON* developed the prototype of a new design for photobio-reactors, which increases the biomass yield appreciably as compared to the usual tubular systems and is currently being optimized at the Anhalt University of Applied Sciences in Köthen.

Production of Chemical Energy Carriers from Microalgae

The algal biomass contains lipids, proteins, and carbohydrates as chemical energy carriers. From them, sustainable liquid energy carriers such as bioethanol and biodiesel may

TAB. 1 | ENERGY EXTRACTION FROM MICROALGAE

Energy Carrier ↓ Energy Content	Manufacturing Process	Development Status	Advantages/Disadvantages Compared to Energy-Yield Plants
Biodiesel ↓ 10.2 kWh/kg ≙ 37 MJ/kg	Esterification	Pilot-plant scale; ~75 % of pilot plants in the USA (*Solazyme, Solex Biofuels*, etc.), 13 % in Europe (*Biofuel Systems*, Spain, etc.)	+ high oil content + suitable fatty-acid profile + comparable quality to fossil sources – high investment costs – high manufacturing costs
Bioethanol ↓ 7.4 kWh/kg ≙ 26.8 MJ/kg	Fermentation	Currently no applications	– high manufacturing costs – low process efficiency
	Extracellular product concentration	Pilot plants in the USA and Mexico (*Algenol*)	+ high process efficiency + low costs – poor long-term stability
Biohydrogen ↓ 2.75 kWh/m^3 ≙ 9.9 MJ/m^3	Oxygen-free intracellular production	Research/laboratory scale	+ simple apparatus technology + low investment costs – low productivity
Biomethane ↓ 10 kWh/m^3 ≙ 36 MJ/m^3 (up to 16 MJ/kg of dry biomass)	Fermentation	Research; pilot plants in Narbonne, France (open pond, fermentation of the whole algal biomass)	+ high efficiency + technology and plants available + minimal investment costs + waste algal biomass suitable as substrate + wet biomass can be utilized + recycling of CO_2 from biogas to the algae systems is possible

be manufactured, as well as gaseous products like methane or hydrogen (see Table 1 and Fig. 2).

Current research is focused on biodiesel production. It is being supported in particular in the USA with several 100 million $, which has led to the founding of ca. 200 firms. The air transportaton industry relies on liquid fuels, due to their high energy densities. It is therefore developing an aircraft fuel from algae, which is planned to replace kerosene in a stepwise process. In Germany, owing to the available infrastructure, work is being carried out intensively on the fermentation of algal biomass to biogas.

Biodiesel and Kerosene Substitutes

Many species of microalgae store fats as reserve materials (triacyl glycerides, a major component of lipids), from which biodiesel can be obtained. Under normal conditions, they produce 15–30 % of their dry weight as lipids. Under stress or limited nutrition, they yield larger quantities of fats, which can be converted into biodiesel by esterification with methanol. If their supply of nitrogen is limited, for example, the algae can no longer synthesize proteins. Since they still try to store their converted light energy, they instead make more fats. In order to obtain biodiesel of the desired quality (DIN EN 14214) from the algae, one requires exact knowledge of the influence of the cultivation parameters (composition of the medium, pH value, temperature, light intensity). These have a strong effect on the profile of the fatty acids formed.

The potential of many species of microalgae to produce oil has been the subject of laboratory research [4]. Selected species are currently under investigation in various pilot and production plants. In June, 2010, the EADS concern demonstrated the first flight of an aircraft powered by algal fuel. In spite of its higher energy content than kerosene, it cannot compete with the latter in terms of cost. For that, the area-specific investment costs, the requirements for process energy and the efforts needed for conditioning are still too great. Here, further research is necessary to find better technical solutions.

Considered purely from the viewpoint of energy, biodiesel from microalgae is attractive. The algae could theoretically achieve a PCE value of 8–10 % and could thus produce biomass energy at a rate of up to 8,030 GJ ha^{-1} a^{-1}. Such a value has, to be sure, not yet been achieved in practice. Depending on the oil content of the algal biomass, specific energy contents between 18.0 megajoules per kg (MJ kg^{-1}) for an oil-free biomass and 37.9 MJ kg^{-1} for pure algal oil can be obtained. If an increase of the PCE from currently 2–5 % to 6.5 % can be achieved, then at an oil content of 25 %, yields of nearly 230 tons of biomass per hectare and year, and over 60,000 liters of oil are possible [5] under proper conditions of climate and insolation. For rapeseed, the yield is currently not quite 1,200 liters of oil per hectare and year. Microalgae could thus permit a considerably higher area-specific yield than conventional energy-yield plants.

FIG. 2 | AN ALGAL BIO-REFINERY

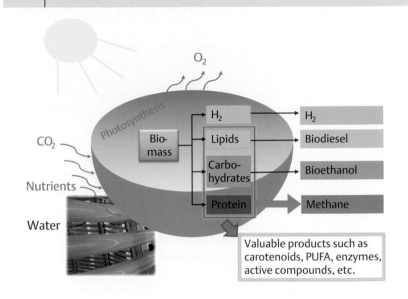

However, up to now there has been no computation of the net energy balance. Furthermore, the overall costs are high. Present estimates assume that using current technology, for the production of 1 kg of biodiesel from algae with an energy yield of 40 MJ/kg (25 % lipid content), roughly 152 MJ of energy must be invested. The single step of separating out the fats consumes 85 % of this process energy requirement [6]. Much research is still needed.

Bioethanol

Many species of macroalgae, but also some microalgae, store the energy of sunlight in their cells in the form of carbohydrates. These can be fermented to give ethanol, i.e. alcohol. However, for currently available processes, the carbohydrates must first be removed from the cells. This decreases the efficiency and increases the cost; thus, these processes have not been commercially successful.

At present, genetically optimized cyanobacteria are being utilized to extract the intracellular alcohol formed from within the cells and into the liquid medium, so that the cost-intensive product separation step can be dispensed with. The first pilot plants for ethanol production in this manner have been built in the USA and Mexico.

Increasingly, macroalgae are being used for the production of bioalcohols. They grow rapidly and can be harvested up to six times a year. Brown algae contain up to 55 % of various sugars, which after extraction can be converted to alcohols.

Biogas and Hydrogen

The fermentation of algal biomass (from micro- and macroalgae) to biogas, a mixture of methane, carbon dioxide and trace gases, is being investigated on a laboratory scale and in the first pilot plants. Currently, methane yields of 0.2 (2 kWh/kg) to 0.5 m^3/kg (5 kWh/kg) are attainable [2,3].

The methane yield per kilogram of dry mass obtained from algae is thus better than that from the best plant substrates (corn silage: 0.35 m³/kg, i.e. 3.5 kWh/kg). An advantage of this process is the existence of an already established technology in the form of a widespread network of biogas installations. The utilization of the (waste) algal biomass in wet form distinctly reduces the process energy required.

A completely new approach to methane production from algae, without intermediate preparation of biomass, is under investigation at the Universities of Leipzig, Karlsruhe and Bremen. These researchers are aiming to produce biogas directly from CO_2 in a two-chambered solar module. Compared to classical biomass fermentation, the area-specific yields are higher by a factor of ten. Here, an algal biofilm in the first chamber, using a special process, converts the products of photosynthesis into a substance (glycolates) which can be used by methane bacteria in the second chamber to produce usable methane directly [7].

Microalgae can also yield hydrogen photobiologically [8], if they are kept on a restricted diet. The basis for this process is likewise an intervention into the photosynthetic system. Hydrogen production from algae has the advantage that, as in ethanol production, no energy-intensive solid-liquid separation steps are required. Disadvantages are however the high sensitivity of the process towards oxygen and its low productivity.

Conclusions

Energy conversion using photovoltaics or wind power plants is currently more efficient than using algae to produce biofuels. However, in addition to fuels, a number of organic compounds can be obtained from the algae, which are currently synthesized starting from petroleum raw materials: Plastics, textiles, dyes, pharmaceuticals, fertilizers and washing agents. Since our resources of petroleum are limited, biotechnology using microalgae must be further developed in Germany to yield sustainable resources. Its potential can be utilized most effectively if optimized photobioreactors are employed to produce valuable substances as coupled products, making use of environmental pollutants such as CO_2 from power plants, or nitrate- and phosphate-containing waste water, while the remaining algal biomass is dedicated to energy-carrier production.

Future production platforms might consist of highly efficient photoreactors in Central and Southern Europe, simple plastic installations in desert areas or brackish swamp regions in the tropics, or floating basins in nutrient-poor ocean bays. Microalgae have proved their potential, and the technical developments are well on their way to realize it in an economically and environmentally suitable manner.

Summary

CO_2-neutral fuels can be obtained from algal biomass: Bioethanol, biodiesel, methane and hydrogen. Thanks to the high efficiency of their metabolisms, algae can yield higher biomass harvests per surface area and time than the energy-yield plants that have thus far been used. They require no arable lands and can make use of CO_2 which would otherwise be harmful to the climate. The key to sustainable biofuel production lies in minimizing all the costs and simultaneously maximizing the productivity of microalgae, including them in a bio-refinery concept.

References

[1] B. Fott, *Algenkunde*, Gustav Fischer Verlag, Stuttgart **1971**.
[2] A.S. Carlsson *et al.*, Micro- and Macroalgae: Utility for Industrial Applications, Outputs from the EPOBIO Project, York **2007**.
[3] K.J. Hennenberg *et al.*, Aquatic Biomass: Sustainable Bioenergy from Algae?, Öko-Institute e.V., Darmstadt **2009**.
[4] C. Griehl *et al.*, Microalgae Growth and Fatty-Acid Composition Depending on Carbon Dioxide Concentration, in *Microalgae: Biotechnology, Microbiology and Energy* (Ed.: M.N.- Johansen), Nova Science Publishers, New York **2011**, 413.
[5] E. Stephens *et al.*, Nat. Biotechnol. **2010**, *28*, 126.
[6] H.H. Koo *et al.*, Bioresource Technology, **2007**, *102*, 5800.
[7] B. Kaltwasser and T. Gabrielczyk, transcript **2011**, *17 (6)*, 60.
[8] F. Lehr *et al.*, J. Biotechnol., **2012**, in press.

About the Authors

Carola Griehl (left) is professor of biochemistry at the Anhalt University of Applied Sciences in Köthen, and since 2004 vice president and vice director of the Life Science Center there. Her research focuses on the production of lipids and active agents from algae and the fermentation of algal waste biomass and protein-rich wastes to yield methane. *Simone Bieler* (center) is a research assistent in the research group of Carola Griehl. *Clemens Posten* (right) is professor for bioprocess engineering at the Karlsruhe Institute of Technology (KIT). His research area is in photo-biotechnology, with a focus on reactor design and process development.

Address:
Prof. Dr. Carola Griehl, Anhalt University of Applied Sciences Department of Applied Biosciences and Process Engineering, Bernburger Strasse 55, 06366 Köthen, Germany.
c.griehl@bwp.hs-anhalt.de

The Karlsruhe bioliq® Process

Synthetic Fuels from the Biomass

BY NICOLAUS DAHMEN | ECKHARD DINJUS | EDMUND HENRICH

Biofuels could replace a portion of the currently-used fossil energy carriers in the near term. To make this possible, raw materials produced over wide-spread areas would have to be made accessible to industrial users of fuels and chemical raw materials on a large scale. The two-stage gasification concept bioliq® offers a solution to this problem.

Fossil energy carriers form the basis of today's energy supplies. Even though predictions of the time remaining until they are completely exhausted differ widely, there can be no doubt that they will in the long run be used up. As current trends in the prices for petroleum and natural gas on the world market demonstrate, even minor perturbations on a global scale can occasionally produce serious price rises with corresponding negative effects on the world's economy. A consistent utilization of renewable energy sources would alleviate these uncertainties and would at the same time contribute to a reduction of CO_2 emissions into the atmosphere.

While hydroelectric power, geothermal heat, solar energy and wind power are suitable primarily for the production of electric power and for space heating, the biomass, uniquely among sustainable carbon sources, can play an important role in the production of motor and heating fuels as well as of organic starting materials for chemical synthesis.

Biogenic fuels can – even in the short term – replace a portion of the present fossil energy sources and thereby make a contribution to the reduction of CO_2 emissions in the transportation sector. The aspects mentioned above, together with existing legal and economic requirements, are contributing to increased political and economic pressure to search for solutions. For example, the EU requires in a recent Directive that the fraction of biogenic motor fuels of the total used must be increased to 10 % by the year 2020. The major portion of this should consist of biofuels whose potential for reducing CO_2 emissions (i.e. savings on carbon emissions in comparison to fossil fuels) amounts to at least 60 % (currently 35 %).

Such a potential for reduction of CO_2 emissions is characteristic of the fully synthetic biogenic fuels of the second generation, also known as BtL fuels (Biomass-to-Liquid). They can be produced using a broad palette of possible raw materials and even employing whole plants. These can be agricultural and forest residues such as straw, waste forest wood, or all the other dry biomass, including energy-yield plants. BtL fuels have the advantage that they are purer and more environmentally friendly than petroleum-based fuels. Furthermore, they can be adjusted to meet special requirements, for example from the automobile manufacturers, and the ever stricter exhaust emission norms. They can be directly utilized within the distribution infrastructure which is already in place today, requiring no new engine technology, and they support the same vehicle operating ranges as petroleum-based motor fuels.

Obstacles to the Use of the Biomass

In comparison to the use of fossil energy carriers, the production of synthesis gas from the biomass is more complex and more expensive. Therre are several hurdles to the technical application of biomass fuels on a large scale.

For one thing, the biomass accumulates over widespread geographical areas and therefore has to be collected and transported, often over long distances. In particular, less valuable biomass such as straw or forest wood residues have a low volumetric energy density (baled straw ca. 2 GJ/m³, as compared to 36 GJ/m³ for diesel fuel). Here, the question arises as to the distances over which it is economically and energetically reasonable to transport these materials. In addition, there is a great variety of potentially usable biomass materials. The processes employed must guarantee the utilization of the largest possible bandwidth of raw materials. Biomass furthermore consists of heterogeneous solid fuels with to some extent differing chemical compositions; and solid fuels in principle require a greater processing effort.

Making use of already established technologies, e.g. for the processing of fossil-fuel raw materials, helps to shorten the development phase and reduce risks. In particular, the

Renewable Energy. Edited by R. Wengenmayr, Th. Bührke. Copyright © 2013 WILEY-VCH Verlag GmbH & Co. KGaA, Weinheim

FIG. 1 | CONCEPT

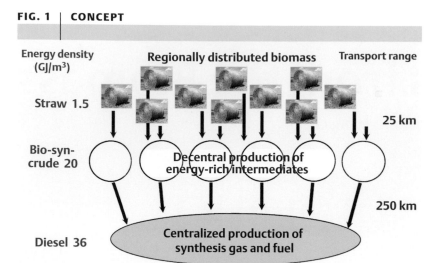

Energy density (GJ/m³)

Regionally distributed biomass

Transport range

Straw 1.5

25 km

Bio-syn-crude 20

Decentral production of energy-rich intermediates

250 km

Diesel 36

Centralized production of synthesis gas and fuel

The decentral-centrally deployable bioslurry gasification concept envisages the production of an energy-rich inter-mediate, which can be economically transported over longer distances, and can then be converted to synthesis gas and fuels in large, centralized installations.

high ash content of many biomass materials causes problems for thermochemical processes, for example due to corrosion or agglutination and blockage of the apparatus. 'Ash content' refers here to the proportion of salts and minerals present.

Motor-fuel synthesis requires a tar-free, low-methane synthesis gas at high pressures of 30 to 80 bar, and the laborious elimination of trace impurities which would act as catalyst poisons. On the other hand, this allows, or at least

FIG. 2 | PROCESS STEPS

Straw → Rapid pyrolysis

Bio-syncrude production

Decentral

High-pressure flow gasification

Gas purification and conditioning

Central

Fuel production → **Synthetic fuels**

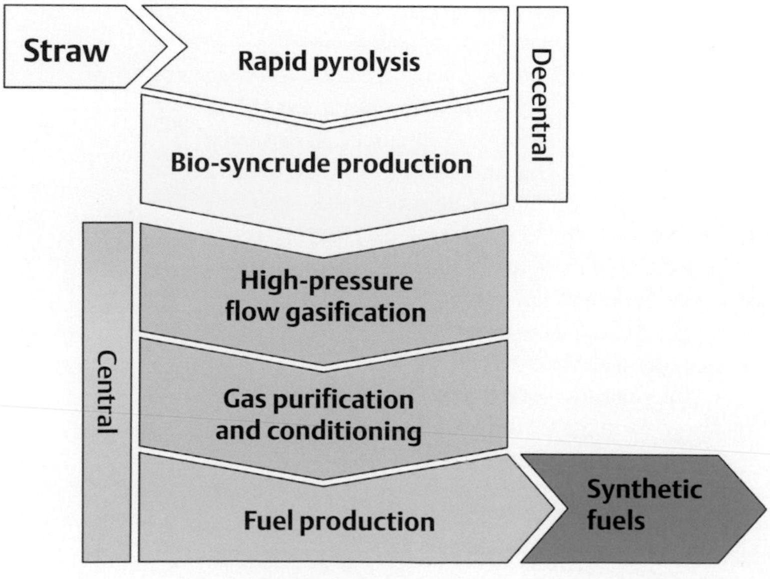

The process steps in the bioliq® Process.

facilitates, meeting the stricter exhaust emission norms when the fuel is burned.

Biomass materials, with their average chemical composition of $C_6H_9O_4$, yield on gasification a C/H ratio near to one, which is insufficient for the production of hydrocarbons. This necessitates an additional process step, the water-gas shift reaction, in which – via addition of water – a part of the CO is converted into hydrogen and CO_2. This leads however to poor carbon efficiency. In the long run, it will be expedient to fill the additional hydrogen requirements by utilizing other sustainable energy sources.

The biomass is taken from the biosphere, and this must in the long term be done in an ecologically compatible manner. Furthermore, the use of the biomass also has socio-economic aspects, since the new role for arable lands, grasslands and forests as providers or processors of energy-yield raw materials requires the establishment of new logistics, income and labor structures.

The Karlsruhe bioliq® Process

At the Karlsruhe Institute of Technology (KIT), we have developed a biomass-to-liquid process intended to overcome these logistical and technical hurdles. The emphasis of our process lies in the use of relatively inexpensive and thus far mostly wasted biomass residues. The Karlsruhe synthetic motor fuel is produced by a process involving several steps, the bioliq® process [4, 5] (Figures 1 and 2).

1. Rapid pyrolysis: In a first step, the decentrally accumulated biomass is converted through rapid pyrolysis into pyrolysis oil and pyrolysis coke (see Figure 3). The air-dried biomass is chopped up and mixed with hot sand, which serves as heating agent, at ambient pressure and under exclusion of air in a double-screw mixing reactor for rapid pyrolysis (Figure 3). The heating, the actual pyrolytic conversion of the biomass particles at about 500° C, and the condensation of the resulting pyrolysis vapors all take place within seconds [6]. Depending on the operating parameters of the reactor and the biomass employed, 40–70 % of a liquid organic condensate (pyrolysis oil) and 15–40 % pyrolysis coke are obtained. The remaining product is a non-condensable pyrolysis gas, whose heat of combustion can be used for preheating the sand or for drying and preheating of the input materials. The mixing reactor used here was developed over 40 years ago by the industry as a "sand cracker" for the rapid pyrolysis of various refinery products [7].

2. Slurry production: The brittle and highly porous pyrolysis coke is mixed with the pyrolysis oil to give a suspension, called bioslurry (Figure 4). For this process, the size distribution of the coke particles is important. Only when their size is sufficiently small is the resulting slurry stable over long periods of time and can be converted rapidly in the following gasification step. The energy density of the slurry relative to its volume is more than an order of magnitude greater than that of dry straw, and this is an advantage for its transport. Rapid pyrolysis is required at this

FIG. 3 | RAPID PYROLYSIS

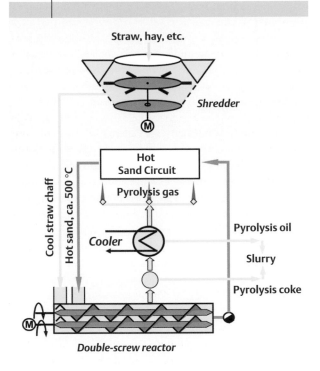

Schematic of the rapid pyrolysis process, which uses a double-screw mixing reactor to produce pyrolysis oil and pyrolysis coke, the precursors of bioslurry.

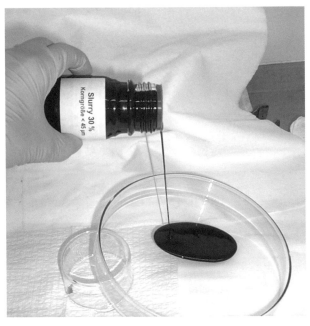

Fig. 4 *The products of the biomass pyrolysis are used to prepare an energy-rich, free-flowing intermediate: bioslurry.*

point in order to obtain the ideal mixing ratio of pyrolysis condensates to pyrolysis coke for the preparation of the bioslurry. This is in turn necessary for a complete utilization of both components [8, 9].

3. Flow gasification: The bioslurry is atomized with hot oxygen in a flow gasifier and is converted at over 1200° C to a tar-free and low-methane crude synthesis gas [10–12]. This flow gasification apparatus is similar to those developed for gasification of Central German salty lignite (soft coal). It is especially well suited for an ash-rich biomass [13]. This is due to a cooling mantle onto which the ash precipitates as molten slag and then drains out of the reactor (Figure 5).

The suitability of this type of gasifier was demonstrated in several test series using different bioslurries and operating parameters in the 3–5 MW pilot gasifier plant at the *Future Energy* company in Freiberg. Bioslurries with up to 33 wt.-% of coke were tested, from which a practically tar-free, low-methane (< 0.1 vol.-%) synthesis gas was obtained. It consists of 43–50 vol.-% carbon monoxide, 20–30 vol.-% hydrogen, and 15–18 vol.-% CO_2. Gasification takes place under a pressure which is dependent on the synthesis to follow, thus avoiding costly compression of the synthesis gas. For example, Fischer-Tropsch synthesis requires pressures up to 30 bar, while the methanol or dimethyl-ether synthesis requires up to 80 bar.

4. Gas purification and conditioning: Before use in a chemical synthesis, the crude synthesis gas must be purified

of particles, alkali salts, H_2S, COS, CS_2, HCl, NH_3, and HCN. This prevents poisoning of the catalysts used in the following synthesis step.

5. Synthesis: The conversion of synthesis gas into motor fuels on a large scale is an established technology. For example, the *Sasol* concern uses the Fischer-Tropsch synthesis to produce more than six million tons of fuel from anthracite coal annually [2]. Natural gas is also increasingly used with this process to yield synthetic products, including diesel fuel and kerosene; for example in the world's largest GtL (Gas-to-Liquid) project PEARL, operated by *Shell* in Qatar.

Methanol production, yielding on the order of many millions of tons per year, is likewise an established process technology. Methanol is on the one hand directly usable itself as a motor fuel. It is used for the synthesis of the anti-knock compound MTBE (methyl tertiary-butyl ether) and for the production of biodiesel through esterification of rapeseed oil, as well as being an input fuel for high-temperature fuel cells. Methanol also serves as an intermediate for a methanol-to-gasoline process, whereby high-octane gasoline is produced with great selectivity [3].

The Current State of Development

Our work up to now in Karlsruhe demonstrates that even bioslurries with a high coke content resulting from biomass pyrolysis products can be completely and safely converted into a tar-free synthesis gas using pure oxygen in a flow gasifier at high pressures. This process is suitable for practically all starting materials which yield a sufficiently stable condensate for suspending the coke powder after rapid pyrolysis.

Now that the technical feasibility of the process has in principle been demonstrated by experiments with our own and with industrial equipment, the overall process is being further developed as rapidly as possible. For this purpose, we are currently setting up a pilot plant in Karlsruhe within the framework of a public grant and with industrial cooperation partners. It will have a biomass throughput of 500 kg/h, and is intended to demonstrate and further develop the process, to show the practicability of the specific process steps, to prepare for scaling-up to a commercially relevant size, and to allow the compilation of reliable cost estimates. The first of three constructional phases, the phase of biomass milling, rapid pyrolysis and continuous mixing of the bioslurry, was completed in 2008. The pyrolysis plant is operated together with the firm *Lurgi AG*, Frankfurt. The 5 MW_{th} gasification plant is also being constructed in cooperation with this firm, and will be completed in the near future, together with the further process steps of gas purification and fuel synthesis. The hot-gas purification stage, developed in cooperation with the *MUT* company from Jena (Germany), consists of a ceramic particle filter, a solid-bed absorber to remove acidic gases (HCl, H_2S) and alkalis from the crude synthesis gas, and a catalytic reactor to decompose organic and nitrogen-containing trace contaminants. We expect clear-cut energy savings from this hot-gas purification in comparison to the usual scrubbing process carried out at low temperatures.

In cooperation with the *CAC* concern from Chemnitz (Germany), we are constructing and operating the two-stage fuel synthesis plant. It utilizes methanol and dimethyl ether (DME) as intermediates and produces high-quality gasoline at a good yield. Unreacted synthesis gas is returned to the process. In 2012, the pilot plant will go into complete operation, including the whole process chain.

Costs and Development Potential

The Karlsruhe Biomass-to-Liquid process is particularly suited to the requirements of the widely distributed biomass production from agriculture: The rapid pyrolysis and production of the bioslurry are carried out at a large number of decentrally located plants. They provide the decisive enhancement in energy density needed for further economical transport of the raw materials. The gasification and the following steps of gas conditioning and synthesis can then be performed at a large central installation of a size which makes it commercially cost-effective, and which is supplied with the bioslurry raw material by road or rail transport.

In this way, using the bioliq process, a ton of synthetic fuel can be produced from about seven tons of air-dried straw. Up to 40 % of the energy originally present in the biomass remains in the liquid product. Byproducts are heat and electric power which can be used to fill the energy requirements of the entire process.

In a possible scenario, about 40 rapid-pyrolysis installations, each with a capacity of 200,000 tons of bioslurry annually, could be set up to supply a central gasification and fuel production plant with an annual capacity of a million tons of fuel. Then, at a price of 70 € per ton for the air-dried starting material, a production price of less than one € per kg of fuel could be realized [15]. With integration of the gasifier into an equipment network of, for example, the chemical industry, the diversification of the usable products can also be broadened. Along with the methanol route, preferred in the Karlsruhe process, and in addition to DME which is under discussion as an alternative liquefied-gas fuel, many oxygen-containing basic chemical products, the so-called oxygenates, can be produced. Ethylene and propylene, the starting materials for roughly half of the world's plastics production, can also be synthesized in this way. Besides the option of the utilization of biomass as a source of carbon, necessary in the long term, economically favorable processes could also be developed in the foreseeable future.

The focus for the process development is currently on low-grade biomass, which thus far has not been used at all, such as surplus grain straw, barn straw or waste wood. The use of solid wood is not seen as a fruitful solution in the long term. Even though this less problematic starting material might permit the technical realization of the process to be achieved more rapidly, it can be expected that the demand for solid wood will increase due to its uses in construction, for cellulose manufacture, and for decentralized and household heat and energy production.

FIG. 5 | THE FLOW GASIFIER

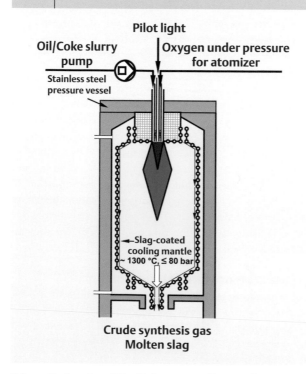

Pilot light

Oil/Coke slurry pump

Oxygen under pressure for atomizer

Stainless steel pressure vessel

← Slag-coated cooling mantle ~ 1300 °C, ≤ 80 bar

Crude synthesis gas
Molten slag

Schematic drawing of the high-pressure flow gasifier, in which the bioslurry is converted to synthesis gas using pure oxygen at temperatures above 1200 °C.

The accompanying systems analysis research [1,14] leads us to expect an annual production of about 5 million tons of synthetic motor fuel from the use of waste forest wood and surplus straw alone, together about 30 million tons of dry starting material. This corresponds to roughly 10 % of the current consumption of petrol and diesel fuels in Germany [15]. Combined with other biochemical and physico-chemical processes, a still higher-quality utilization of the biomass, in the form of a biomass refinery, should be feasible. Similarly to today's petroleum refineries, it would use a broad spectrum of raw materials to produce a variety of basic chemical products and fine chemicals, making use of synergy effects to increase economic productivity, which would result in a clear-cut reduction of the consumption of fossil resources by the chemical industry.

Acknowledgments

For support of the bioliq® pilot plant, we thank the German Federal Ministry for Nutrition, Agriculture and Consumer Protection (BMELV), the Agency for Sustainable Raw Materials (FNR), the State of Baden-Württemberg, and the European Union.

References

[1] L. Leible et al., FZK-Nachrichten **2004**, 36, 206.

[2] R.L. Espinoza et al., Applied Catalysis A: General **1999**, 186, 13 and 41.

[3] W. Liebner and M. Wagner, Erdöl, Erdgas, Kohle **2004**, 120, 323.

[4] N. Dahmen and E. Dinjus, Chemie Ingenieur Technik **2010**, 82, 1147.

[5] N. Dahmen and E. Dinjus, MTZ **2010**, 71, 864.

[6] C. Kornmayer et al., DGMK Tagungsbericht **2006**-2, 185.

[7] R.W. Rammler, Oil and Gas Journal **1981**, Nov. 9, 291.

[8] K. Raffelt, E. Henrich, and J. Steinhardt, DGMK Tagungsbericht **2004**-1, 333.

[9] K. Raffelt et al., DGMK Tagungsbericht **2006**-2, 121.

[10] E. Henrich, E. Dinjus, and D. Meier, DGMK Tagungsbericht **2004**-1, 105.

[11] M. Schingnitz and D. Volkmann, DGMK Tagungsbericht **2004**-1, 29.

[12] M. Schingnitz, Chemie Ingenieur Technik **2002**, 74, 976.

[13] M. Schingnitz et al., Fuel Processing Technology **1987**, 16, 289.

[14] L. Leible et al., DGMK Tagungsbericht **2006**-2, 23.

[15] E. Henrich, N. Dahmen and E. Dinjus, Biofuels Bioprod. Bioref. **2009**, 3, 28.

Summary

Synthetic fuels from the biomass can provide an important contribution to a renewable energy economy. The Karlsruhe BtL concept bioliq® aims at bringing decentral production in line with centralized processing on an industrial scale. To this end, thermochemical methods are employed: rapid pyrolysis for the production of a readily transportable, energy-rich intermediate product, and entrained-flow gasification to yield synthesis gas and to process it further into the desired fuels. The bioliq process was distinguished with the BlueSky Award by the UN Organization UNIDO in 2006.

About the Authors

Nicolaus Dahmen studied chemistry at the Ruhr University in Bochum, obtained his doctoral degree in 1992 and moved in the same year to the Karlsruhe Research Center, now KIT. There, he is currently concerned with the thermochemical transformation of biomass into gaseous, liquid and solid fuels. As project leader, he is responsible for the construction of the bioliq® pilot plant. In 2010, he obtained his Habilitation at the University of Heidelberg.

Eckhard Dinjus began his studies of chemistry in 1963 at the Friedrich Schiller University in Jena and completed his doctorate there in 1973. In 1989, he obtained the Habilitation, and thereafter he was leader of the Research group " CO₂ Chemistry" of the Max Planck Society. Since 1996, he has been director of the Institut for Technical Chemistry, now IKFT at KIT, and he occupies the chair of the same name at the University of Heidelberg.

Edmund Henrich studied chemistry at the Universities of Mainz and Heidelberg. He received his doctorate in Heidelberg in 1971 and the Habilitation in 1993 in the field of radiochemistry. He has worked at the Karlsruhe Research Center since 1974, and as Division Leader of the Institute for Technical Chemistry there, he is responsible for R and D activities relating to the Karlsruhe BtL process. Since 2005, he has been extraordinary professor at the University of Heidelberg, and is professor emeritus there since 2009.

Address:

Dr. Nicolaus Dahmen, Dr. Eckhardt Dinjus, Dr. Edmund Henrich, Karlsruhe Institute for Technology (KIT), Institut fuer Katalyseforschung und -technologie (IKFT), Postfach 3640, 76021 Karlsruhe, Germany. nicolaus.dahmen@kit.edu

The Solar Updraft Tower Power Plant
Electric Power from Hot Air

BY JÖRG SCHLAICH | RUDOLF BERGERMANN | GERHARD WEINREBE

A solar updraft tower – or solar chimney – power plant combines the greenhouse effect with the chimney effect, in order to obtain electrical energy directly from solar radiation. The plants must be very large, with towers 700 m and higher, to produce energy in an economically competetive fashion.

From earliest times, humans have made active use of solar energy: Greenhouses aid in the cultivation of food crops, the chimney updraft principle was employed for ventilation and cooling of buildings, and the windmill for grinding grain and pumping water. The three essential components of a solar updraft tower power plant – a hot air collector, updraft tower, and wind turbines – have thus long been known and used. In the solar thermal updraft tower power plant, they are simply combined in a new way (Figure 1).

Around 1500, Leonardo da Vinci sketched an apparatus which made use of the rising warm air in a chimney to turn a spit for roasting meat. Already in 1903, the Spaniard Isidoro Cabanyes described a motor powered by updraft currents in his article "Project for a Solar Motor" in the journal "La Energía Eléctrica" [1]. The modern combination with a generator for producing electric power was first described by Hanns Günther more than seventy years ago [2]. In the 1960's and 70's, the Frenchman Edgar Henri Nazare and the German Michael Simon worked on this topic. Nazare, however, intended to use a large tower to produce an artificial cyclone, without a collector. We took up the idea of an updraft tower with a collector and developed it further to its present-day form and to technical maturity.

After initial trials and experiments in a wind tunnel, in 1981/82 we were able to set up an experimental installation in Spain. It operated successfully over a period of seven years. Since then, we have continued in our efforts to construct a commercial solar updraft tower power plant with high output power, thus far without success.

INTERNET

Schlaich Bergermann Solar
www.solar-updraft-tower.com

Information about Edgar Nazare
bit.ly/qPcWKu
bit.ly/nYsPVA

Operating Principles

The principle of the solar updraft tower power plant is shown in Figure 1 (an animation of the operating principle can be viewed in [3]). Under a transparent roof, which is flat, circular, and open around its outer perimeter, the air is warmed by solar radiation (greenhouse effect); together with the natural ground below, the roof forms a hot air collector. At its center is a vertical updraft tower with large air inlets around its base. The roof is attached to the base of the tower by an airtight seam. Since hot air has a lower density than cool air, it rises in the updraft tower. The resulting draft pulls more hot air in from the collector, and at its perimeter, cooler air flows in. The solar radiation thus powers a continuous updraft in the updraft tower. The energy contained in this air flow is converted to mechanical energy with the aid of pressure-staged air turbines which are mounted at the base of the updraft tower, and finally, using generators coupled to the turbines, it is converted to electrical energy.

The solar thermal updraft tower power plant is technically very similar to a hydroelectric plant – thus far the most successful type of power plant making use of sustainable energy sources: The collector roof corresponds to the water reservoir, the updraft tower to the penstocks. Both types of power plants utilize pressure-staged turbines, and both have low power-generating costs due to their extremely long lifetimes and their low operating expenses. The collector roof and the reservoir require comparable areas for the same power output. The collector roof, however, can be set up in arid desert regions and readily removed after its useful life, while as a rule, the reservoirs of power dams flood living (and often even inhabited) land.

Continuous 24-hour operation can be achieved by using water-filled tubes or sacks which are laid out on the ground under the collector roof. The water is heated during the day and releases its heat to the air at night (Figure 2). The tubes need to be filled only once; there is no further water consumption. Thus, solar radiation can produce a continuous rising wind in the tower [4].

In order to mathematically describe the time-dependent production of electrical energy from a solar updraft tower power plant, a comprehensive thermodynamic and aero-

FIG. 1 | AN UPDRAFT TOWER POWER

FIG. 1 | AN UPDRAFT TOWER POWER

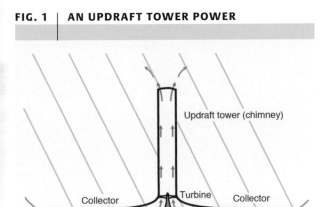

Operating principle of the solar updraft tower power plant.

FIG. 2 | HEAT STORAGE PLANT

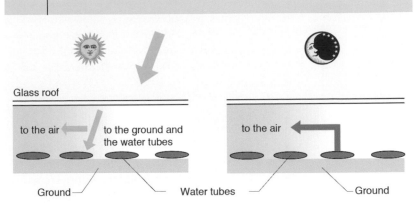

Heat storage using water-filled tubes.

dynamic model is needed [5]. A good description of the thermodynamics of the solar updraft tower power plant as a cyclic process is to be found in [6]. In the following, we explain the basic relations in simplified form.

Generally speaking, the output power P of a solar updraft tower power plant can be computed as the product of the solar energy input, $\dot{Q}_{solar}$, multiplied by the various efficiencies η of the collector, the updraft tower, and the turbine:

$$P = \dot{Q}_{solar} \cdot \eta_{plant} = \dot{Q}_{solar} \cdot \eta_{coll} \cdot \eta_{tower} \cdot \eta_{turbine}. \qquad (1)$$

The solar energy input into the system can be written as the product of the solar global radiation G_h onto a horizontal area, multiplied by the area of the collector, A_{coll}:

$$\dot{Q}_{solar} = G_h \cdot A_{coll}. \qquad (2)$$

The tower converts the heat current delivered by the collector into mechanical energy. This consists of the kinetic energy of the convective flow and potential energy; the latter corresponds to the pressure drop at the turbine. The density difference of the warm and cool air thus acts as the driving force. The column of lighter air in the tower is connected to the surrounding atmosphere at the base of the tower and at its top, and therefore experiences a lift or updraft force. A pressure difference Δp_{tot} is established between the base of the tower and its surroundings:

$$\Delta p_{tot} = g \cdot \int_0^{H_t} (\rho_a - \rho_t) dH. \qquad (3)$$

Here, g is the acceleration of gravity, H_t the height of the tower, ρ_a the density of the surrounding air, and ρ_t the density of the air in the tower. Thus, Δp_{tot} increases proportionally to the height of the tower.

The pressure difference Δp_{tot} can be decomposed into a static component Δp_s and a dynamic component Δp_d:

$$\Delta p_{tot} = \Delta p_s + \Delta p_d. \qquad (4)$$

Frictional losses have been neglected here. The static pressure difference equals the pressure drop at the turbine; the dynamic component describes the kinetic energy of the flowing air.

Knowing the overall pressure difference Δp_{tot} and the volume current of the air in the system, i.e. the product of its mean transport velocity in the tower, c_{tower}, and the tower's cross-sectional area A_{tower}, we can now compute the power contained in the air flow:

$$P_{tot} = \Delta p_{tot} \cdot c_{tower} \cdot A_{tower}. \qquad (5)$$

From this, finally, the thermo-mechanical efficiency of the tower can be given as the quotient of the mechanical power in the flow with the thermal current $\dot{Q}_{tower} = \dot{Q}_{solar} \cdot \eta_{coll}$, which is input into the system:

$$\eta_{tower} = \frac{P_{tot}}{\dot{Q}_{tower}}. \qquad (6)$$

FIG. 3 | OUTPUT POWER

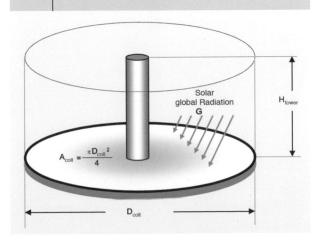

The output power of the solar updraft tower plant is proportional to the collector area and to the height of the tower.

The distribution over static and dynamic components in reality depends on how much energy is withdrawn from the air flow by the turbine. Without a turbine, a maximum flow velocity $c_{tower,max}$ is attained, and the overall pressure difference is converted entirely into kinetic energy, i.e. the air flow is accelerated:

$$P_{tot} = \frac{1}{2} \dot{m} \cdot c_{tower,\,max}^2. \qquad (7)$$

where $\dot{m}$ represents the mass flow rate (mass current) of the air. The flow velocity which results from free convection can be determined by making use of the Boussinesq approximation, which summarizes the temperature-dependent density differences in the air in a "buoyancy term" [7]:

$$v_{tower,\,max} = \sqrt{2 \cdot g \cdot H_{tower} \cdot \frac{\Delta T}{T_0}}, \qquad (8)$$

with T_0 the temperature of the surroundings at ground level, and ΔT the temperature difference between the surrounding temperature and that at the inlet of the tower.

Using equation (6) and the relation for a stationary state, together with (7) and (8), we obtain the tower efficiency:

$$\eta_{tower} = \frac{g \cdot H}{c_p \cdot T_0}. \qquad (9)$$

This simplified description points up one of the fundamental properties of a solar updraft tower power plant: The efficiency of the tower depends only upon its height.

Equations (2) and (9) show that the electrical output power of the solar updraft tower plant is proportional to its collector area and to the height of the tower: It is thus proportional to the volume of the cylinder which contains both collector and tower (Figure 3). Therefore, a desired output power can be obtained either with a high tower and a smaller collector, or with a large collector and a small-er tower. When frictional losses in the collector are taken into account, the linear dependence between the power and the product (collector area × tower height) is no longer strictly valid, especially for collectors of large diameter. Nevertheless, it provides a useful rule of thumb.

The Test Installation in Manzanares

After detailed theoretical preliminary investigations and comprehensive wind-tunnel experiments, in 1981/82 we constructed an experimental installation using funds from the German Federal Ministry of Research and Technology, with 50 kW peak electrical output power. It was located in Manzanares, about 150 km south of Madrid. The test site was put at our disposal by the Spanish energy supplier *Union Electrica Fenosa* (Figure 4) [8, 9].

This research project was intended to verify the theoretical calculations using measured data and to investigate the influence of individual components on the power output and efficiency of the plant under realistic structural and meteorological conditions. To this end, we constructed a tower of 195 m height and 10 m diameter, surrounded by a collector of 240 m diameter. The plant was equipped with extensive measurement and data collection instruments. More than 180 sensors registered the operating parameters of the system at intervals of a few seconds. The main dimensions and some technical data of the plant are set out in Table 1.

The prototype in Manzanares was planned for an operating period of only three years. For that reason, its updraft tower was designed as a sheet-metal tube with support rods, which could be recycled after the end of the experiment. Its wall thickness was only 1.25 mm (!), and it was stiffened every 4 m by external cantilever trusses. The base of the updraft tower was fixed 10 m above ground level onto a ring. This was supported by eight thin pipes, so that the hot air could flow into the updraft tower with almost no hindrance. In order to provide a streamlined transition between the collector roof and the base of the updraft tower, a pre-stressed fabric jacket was installed (Figure 5).

The updraft tower was supported at four levels and in three directions from the foundations by low-cost thin steel

Fig. 4 *The prototype at Manzanares in Spain produced 50 kW of electric power.*

Fig. 5 *The turbine of the prototype plant in Manzanares.*

TAB. 1 | THE PROTOTYPE IN MANZANARES

Tower height	194.6 m
Tower radius	5.08 m
Mean collector radius	122.0 m
Mean roof height	1.85 m
Nominal electric power	50 kW
Plastic membrane collector area	40000 m^2
Glass roof collector area	6000 m^2

rods. Guy wires, which are usual for such a structure, or even a free-standing concrete tower would have been too costly for the limited budget of the project. The sheet-metal tubes were assembled on the ground using an especially developed cyclic lifting procedure. They were lifted in stages using hydraulic jacks while at the same time, the bracing rods were adjusted. This was intended to demonstrate that even high towers can be built by only a few skilled workers. Of course, this intentionally temporary construction is not appropriate for a large updraft tower power plant with a longer planned lifetime. Under realistic conditions, as a rule the updraft tower will be constructed of reinforced concrete.

The collector roof of an updraft tower power plant must not only be transparent to sunlight, but also should have a long operating lifetime. We therefore tried out different plastic sheeting and glass. The experiment was intended to show which material would work best and be most cost-effective on a long-term basis. Glass resisted even violent storms without damage during the operational life of the plant, and proved to be self cleaning; occasional rainfall is sufficient for this. The square plastic membranes were clamped at their edges into profiled channels and were attached at their centers via a plastic plate with a drain opening to the ground. The investment costs for a plastic sheet collector are lower than those for a glass collector roof. However, the sheets became brittle in the course of time and tended to rip. In the meantime, there are more durable plastic materials, which again make plastic sheet collectors a realistic alternative.

After completion of the construction phase in 1982, the experimental phase began: It was to demonstrate that the principle of the updraft tower power plant would function under realistic conditions. It was important for us to obtain data on the efficiency of the newly-developed technology. Furthermore, we wanted to demonstrate that the power plant could be operated reliably in a fully automatic mode. Finally, we wanted to record and analyze its operating behavior and the underlying physical processes on a long-term basis.

Figure 6 shows the important operating data for a typical day: Solar radiation, air flow velocity and electrical power output. These clearly show that for this small plant, without a thermal storage system, the electric power output during the day is closely correlated with the solar radiation input (Figure 7).

FIG. 6 | OPERATIONAL DATA FROM THE TEST INSTALLATION

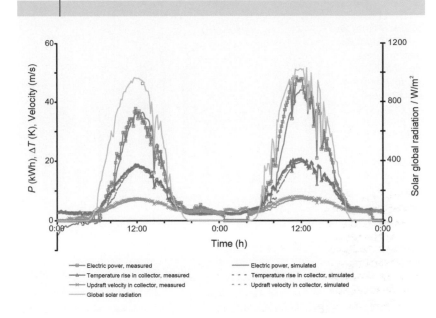

Measured data from Manzanares: Temperature difference, updraft velocity, and electric power on two days (7th and 8th of June, 1987). Measured output energy: 635 kWh; simulated output energy: 626 kWh. No heat storage using water-filled tubes was installed.

However, even at night there is some air flow, so that during some of the night-time hours, power can still be generated (Figure 6). This effect increases with increasing size of the plant (and thus of the collector), i.e. with increasing thermal inertia of the system. This was confirmed for large installations by simulation results.

FIG. 7 | POWER YIELD

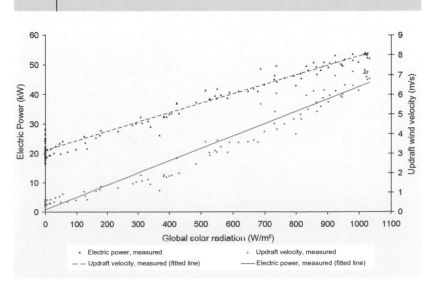

The relation between solar radiation and power for the prototype in Manzanares (8th June 1987).

FIG. 8 | MONTHLY ENERGY PRODUCTION

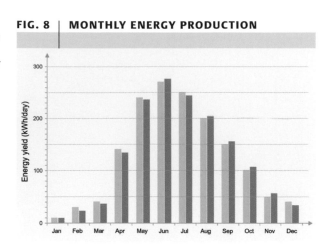

Comparison of the measured (blue) and calculated (by computer simulation, orange) monthly energy production from Manzanares. In the entire year, the plant should yield 44.35 MWh according to the calculation, while the measured value was 44.19 MWh.

During the year 1987, the plant was in operation for 3197 h, corresponding to an average daily operational period of 8.8 h. As soon as the air flow velocity surpassed a certain value – typically 2.5 m/s – then plant operation started automatically and synchronized with the power grid. The installation as a whole and the individual components operated very reliably.

From the data, we developed a model for a computer simulation. We wanted to obtain a solid understanding of the physical processes and to identify possible starting points for improve-ments with the aid of the simulations. The computer model describes the individual components, their performance and their dynamic interactions. It is based on a finite-element approach and takes the conservation equations for energy, momentum and mass into account. At present, it is a developmental tool which considers all the known relevant physical effects. Using it, the thermodynamic behavior of large solar updraft tower power plants under given weather conditions can be reproduced [10, 11].

Figure 8 gives a comparison between the average monthly energy output as calculated by the simulation and the measured value from the test plant. The two values agree very well.

In summary, we can say that the thermodynamic processes in a solar thermal updraft tower power plant are well understood. The computational models permit a realistic simulation of the operational behavior of the plant under given meteorological conditions.

Large Power Plants

Our detailed investigations, supported by extensive wind-tunnel experiments, show the following: The thermodynamic calculations for the collector, the tower and the turbine can be reliably scaled up to a larger magnitude. The small pilot plant in Manzanares covered a much smaller area and contained a much lower volume than for example the 200 MW plant which we will introduce below. Nevertheless, the thermodynamic characteristics of the two plants are astonishingly similar. If we consider for example the temperature increase and the flow velocity in the collector, then in Manzanares, we measured up to 19 K and 12 m/s, while the simulation of a 200 MW plant yields average values of 18 K and 11 m/s.

Such comparisons support the premise that we can use the measured data from Manzanares and our updraft tower plant simulation program in order to design large-scale plants. In Figure 9, the results of a simulation for a site in Australia are given. For each season, a period of four days is simulated. This plant is assumed to have an additional heat-storage facility and operates around the clock, in particular also in the autumn and winter – then of course with a reduced power output.

Today, more efficient materials which can be installed on a more cost-effective basis are available for various collector designs; these can be either of glass or plastic.

Large plants would require towers of up to 1000 m height. These are a construction challenge, but they are within the realm of possibility today. The CN Tower in Toronto rises to a height of nearly 600 m, and the skyscraper Burj Dubai is over 800 m. In contrast to a skyscraper, the tower of a solar updraft tower power plant need be only a simple hollow cylinder. It is not particularly slim, and will thus stand securely, while the technical and structural requirements are considerably less strict than those for an inhabited building.

There are various different methods for constructing such a tower: free-standing reinforced concrete tubes, steel sheet tubes supported by guy wires, or cable-net construction with a cladding of sheet metal or plastic membranes.

FIG. 9 | A 200 MW UPDRAFT TOWER POWER PLANT

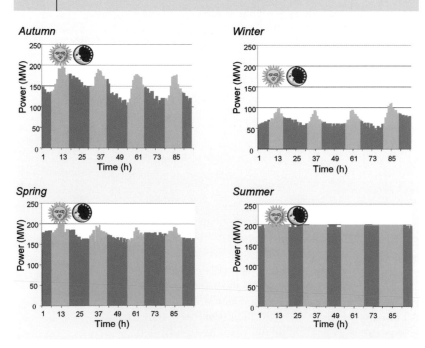

Simulation results for a solar updraft tower power plant with 200 MW output power, showing the peak-demand output with additional thermal storage capacity. The simulation is for a site far from the equator, with strong seasonal variations.

The design procedures for such structures are all well established and have already been utilized for cooling towers; thus, no new developments are required. Detailed static and structural-mechanical investigations have shown that it would be expedient to stiffen the tower in several stages, so that a relatively thin wall thickness would suffice. One solution is to use bundles of thin wires in the form of "flat" spoked wheels which span the cross-sectional area of the tower (Figure 10). This is perhaps the only real novelty in updraft tower power plants as compared to existing structures.

For the dimensioning of the machines, we can fall back on experience from water and wind power plants, cooling-tower technology, wind-tunnel fans – and of course from the prototype plant in Manzanares. Initially, a single, large vertical-shaft turbine in the updraft tower seemed to be the most obvious solution, as suggested in Figure 1 and used in the Manzanares plant (Figure 5). Newer cost estimates have in the meantime convinced us to plan in newer designs for a larger number of horizontal-shaft turbines. These form a ring at the base of the updraft tower, at the point where the collector is attached to the tower. This allows the use of smaller turbines, which are considerably less expensive. Furthermore, their redundance guarantees a high availability, since when individual turbines are out of service, the others can continue to generate power. In addition, reduced-output operation is improved, since the plant's power output can be controlled by switching the individual turbines on or off.

The energy yield of an updraft tower power plant is proportional to the global insolation, its collector area and the height of its updraft tower. There is thus no physically optimal size, only an economic optimum: The best dimensions for the given site are found in terms of the specific component costs for a suitable collector, tower, and turbines. Thus, plants of different dimensions will be built for different sites – in each case at minimal cost: If collector area is cheap but reinforced concrete is expensive, then one would construct a large collector and a comparatively small tower; and if the collector area is expensive, then one would build a smaller collector and a larger tower.

Table 2 gives an overview of the typical dimensions of solar updraft tower power plants. The numbers are based on typical international materials and construction costs. The costs of unskilled labor have been taken here to be 5 ¢/h; we are thus considering a site in a developing or emerging nation. The updraft tower power plant is especially suitable for such countries. In particular, it becomes clear that the power-generating costs decrease significantly with increasing plant size. In order to achieve economically viable operation, a solar updraft tower power plant must therefore be of a certain minimum size.

The Mildura Project

In Australia, there are not only large, flat unused open spaces with a high insolation, but also reserves of the fossil energy carriers, coal and natural gas. Up to late 2007, the Australian government was banking almost exclusively on the use of coal for its future energy supply. However, in 2001, it established a political framework in terms of the Mandatory Renewable Energy Target, which provides a protected, preferred market for sustainable energy sources. Therefore, the initial situation for construction of a solar updraft tower power plant was relatively favorable.

The enterprise *EnviroMission* was founded with the goal of marketing the solar updraft tower power plant technology of *Schlaich Bergermann Solar* (Stuttgart) in Australia. Initially, we in Stuttgart were commissioned to prepare a detailed market study, which was successfully completed. Independently of it, the technical feasibility was also affirmed by the well-known Australian Engineering firm *Sinclair Knight Merz*. Not least as a result of these encouraging results, the Australian construction company *Leighton Contractors*, a subsidiary of the German *HochTief*, decided to join the project. After the planning of the construction of the tower and the collector, carried out in cooperation by *EnviroMission* and *Leighton Contractors*, the latter confirmed their constructability and submitted an estimate for the project.

The planned power plant would have a tower of 1000 m height and a collector diameter of 7000 m (Figure 10). Its overall electric output power of 200 MW would be divided among 32 turbines, each with a nominal output power of 6.25 MW.

Unfortunately, it became clear in the end that the expected earnings were not high enough to provide potential investors with a sufficiently high profit margin in relation to the risks of building the first commercial updraft power plant: the financing was unsuccessful.

Fig. 10 *A view from above of a projected large-scale solar updraft tower power plant. The spokes of the internal bracing in the tower, and the ring-shaped visitors' platform with its elevator are clearly visible* (graphics: Schlaich Bergermann Solar).

TAB. 2 | TYPICAL BENCHMARK DATA FOR SOLAR UPDRAFT TOWER POWER PLANTS

Nominal output power	MW	5	30	100	200
Tower height	m	550	750	1000	1000
Tower diameter	m	45	70	110	120
Collector diameter	m	1250	2950	4300	7000
Energy output[A]	GWh/a	14	87	320	680
Energy generating cost[B]	€/kWh	0.34	0.19	0.13	0.10

[A] At a site with a total global insolation of 2300 kWh/(m²a).
[B] With a linear amortisation over 20 years and an interest rate of 6 %.

Outlook

In the meantime, together with project developers, we are searching for possibilities to construct a plant of medium size with about 30 MW output power at a site with particularly favorable boundary conditions. The advantage of such a project is that, although the power generating cost is notably higher than for a large, 200 MW plant, the necessary investment, and thus the financial risk, is also clearly lower. In addition, a medium-sized updraft tower power plant has other advantages: It is considerably simpler to find a suitable site, the constructability of the tower will not be questioned by the public, since comparably large structures already exist, and the percentage of potential profits which would be contributed by supplementary income sources such as tourism and naming rights is greater. Independently of this, the long-term goal must be the construction of large power plants with output power in the 200 MW range, since only then, leaving government subsidies out of consideration, will it be possible to establish sustainable, ecologically compatible and at the same time economically feasible power generation.

This challenge must now be met. Once a solar updraft tower power plant has been built and is delivering power to the grid, additional plants will no doubt follow rapidly. After all, they offer many advantages, since their construction does not involve *consumption* of resources, but simply their *commitment* [4]. Solar updraft tower power plants are constructed essentially from concrete and glass, that is from sand and (self-generated) energy. They can therefore reproduce themselves even in the desert – a truly sustainable energy source.

Summary

A solar updraft tower power plant combines the greenhouse effect with the chimney effect in order to generate electrical energy directly from solar radiation. Air is heated beneath a glass roof; it then rises through a central tower and drives wind turbines while passing upwards. This simple principle can be successfully put into large-scale operation, as was demonstrated by an experimental plant in Manzanares, Spain. However, solar updraft tower power plants must have enormous dimensions in order to generate electrical current economically. Following the intermediate step of constructing and operating a plant of medium size, with e.g. 30 MW of output power, the goal is to construct large-scale plants, with tower heights of 1000 m, collectors 7 km in diameter, and an output power of 200 MW.

References

[1] I. Cabanyes, La Energía Eléctrica, **1903**, 5, 4 (in Spanish); excerpts are available online at: www.fotovoltaica.com/chimenea.pdf.
[2] H. Günther, In hundert Jahren – Die künftige Energieversorgung der Welt. Kosmos, Franckhsche Verlagshandlung, Stuttgart **1931**.
[3] www.wiley-vch.de/berlin/journals/phiuz/05-05/SBP_Aufwind-kraftwerk_Trailer.wmv.
[4] J. Schlaich et al., Aufwindkraftwerke zur solaren Stromerzeugung, erschwinglich – unerschöpflich – global. Bauwerk-Verlag, Berlin **2004**.
[5] M. A. Dos Santos Bernardes, A. Voß, G. Weinrebe, Solar Energy **2004**, 75 (6), 511.
[6] A. J. Gannon und T. W. v. Backström, Thermal and Technical Analyses of Solar Chimneys, in: Proc. of Solar 2000, (Eds.: J. E. Pacheco, M. D. Thornbloom), ASME, New York **2000**.
[7] H. D. Baehr, Wärme- und Stoffübertragung, 3. edition, Springer, Berlin **1998**.
[8] W. Haaf et al., Solar Energy **1983**, 2, 3.
[9] J. Schlaich et al., Abschlußbericht Aufwindkraftwerk. BMFT-Förderkennzeichen 0324249D, Stuttgart **1990**.
[10] W. Haaf, Solar Energy **1984**, 2, 141.
[11] G. Weinrebe and W. Schiel, Up-Draught Solar Tower and Down-Draught Energy Tower – A Comparison, in: Proceedings of the ISES Solar World Congress 2001. Adelaide (Australia) **2001**.

About the Authors

Rudolf Bergermann, born in Düsseldorf in 1941, studied structural engineering in Stuttgart. From 1974 to 1979, he worked with Schlaich as Senior Engineer for Leonhardt & Andrä, and since 1980, he has been a partner of the agency Schlaich, Bergermann & Partner. He has been chief designer of a number of notable projects such as the Ting-Kau bridge in Hongkong, and holds an honorary doctorate from the University of Cottbus.

Jörg Schlaich, born in 1934, studied architecture and structural engineering at the University of Stuttgart, the TU Berlin and in Cleveland, Ohio. Doctorate 1962. 1963–79 engineer and partner in the firm Leonhardt & Andrä, Stuttgart. 1974–2000 professor and director of the Institute for Solid Structures in Stuttgart. Since 1980, he has been a partner at Schlaich Bergermann & Partner, Stuttgart. Participation in the construction of the Olympic Stadium in Munich. His project "Think" took second place in the competition for the new World Trade Center in 2003.

Gerhard Weinrebe was born in 1965 and studied aeronautics and space technology at the University of Stuttgart. He was a research assistant at the Plataforma Solar de Almería in Spain. He obtained his doctorate from the University of Stuttgart. Since 2000, he has been with Schlaich Bergermann & Partner.

Address:
*Rudolf Bergermann, Schlaich Bergermann Solar, Schwabstraße 43, 70197 Stuttgart, Germany.
r.bergermann@sbp.de*

Tidal-stream Power Plants

Sun, Moon and Earth as Power Source

BY ALBERT RUPRECHT | JOCHEN WEILEPP

Tidal currents can be used in many regions of the earth for generating electric power. The challenge is however to construct robust, corrosion-resistant and reliable plants. We describe them here using the example of the technological developments at Voith Hydro.

This 110 kW pilot plant made by Voith Hydro Ocean Technologies went into operation in early 2011 in Korea (photo: Voith Hydro).

Today, great efforts are being made to harness the energy of ocean currents produced by the tides, for electric power generation. The tides are due essentially to the gravitational attraction of the moon as well as the revolution of the earth around its common center of gravity with the moon. The attraction of the sun modulates this effect, producing two spring tides each month; depending on the local geography, these can have varying strengths from place to place. Together, these effects lead locally to high and low tides. The rise and fall of the water level in turn means that water flows in the direction of the gravitational force, and this produces tidal currents.

Tidal flows thus change their direction, depending on whether the water level is rising or falling. In the free oceans, there are permanent currents, like the Gulf Stream. Whether they can be utilized as energy sources is however questionable, since the consequences for the climate are to a great extent unknown; even a minor change might have major global effects. In the case of tidal currents, this is not a problem, since their kinetic energy is dissipated by friction in any case. Their use for energy conversion thus has at most a very limited, local effect on the environment.

To make use of the energy of tidal streams, a turbine converts the kinetic energy of the water into mechanical energy. A generator attached to the turbine then produces electricity. This distinguishes tidal-stream power plants from tidal power plants; the latter make use of the head of water produced by high and low tides. A tidal plant has for example been operating since the mid-1960's at the mouth of the River Rance near the town of St. Malo in Brittany, France. Such plants back up the water with a dam for a certain pe-

riod of time and make use of the difference in water levels on either side of the dam, i.e. they convert the potential energy of the water back to kinetic energy. In contrast, tidal-stream power plants require no dams, with their major effects on the landscape. Their turbines work in a free stream, like wind turbines, and make direct use of the flowing water's kinetic energy. It is important to note that these two tidal technologies are not in competition for usable sites, but rather complement each other: The one cannot as a rule replace the other, since maximal tidal flow velocities are usually found at different locations from maximal tidal water heights.

Tidal-stream power plants have the advantage of offering a very predictable, sustainable source of energy. They

FIG. 1 | POWER YIELD

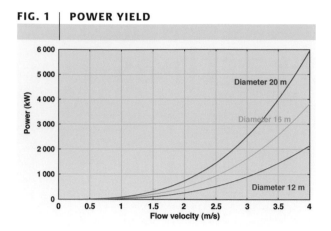

The power yield as a function of the flow velocity.

Concepts for Tidal-Stream Installations

Around the world, there are a number of development and demonstration projects that exhibit a variety of technologies. Essentially, these concepts can be classified as:

- Free-stream, horizontal-axis turbines;
- Free-stream, vertical-axis turbines;
- Resistance rotors; and
- Some few oscillating hydroplane applications, where a hydroplane moves periodically, sweeping out a rectangular area.

Free-stream turbines and oscillating hydrofoil applications make use of the hydrodynamic lift forces on rotating hydrofoil profiles. Resistance rotors (e.g. undershot water wheels) rotate only by virtue of their resistance to the water flow; thus, they have a poor efficiency of less than 50 %. Good free-stream turbines, in contrast, convert up to 85 % of the possible flow energy as in (1) (Betz Limit) into mechanical energy.

Along with free-stream turbines, there are also ducted turbines. The specific power yield relative to the structural area of both types is roughly similar. The corresponding installations differ in the way the turbine is mounted. As with wind-power plants, one can distinguish:

- Gravity-based foundations;
- Monopile foundations; and
- Floating installations.

Which of these concepts is most suitable for a particular site depends upon the nature of the ocean floor and the water depth there.

In recent years, numerous demonstration plants have been placed in the water, and valuable experience has been gained with them. In the meantime, large firms have joined in this development, which is decisive for the realization of industrial-scale tidal-stream parks. Energy providers have also become interested in utilizing this technology on a large scale, so that it is on the threshold of a breakthrough to commercialization.

also have a higher energy density than e.g. wind plants. On the other hand, the installations must survive in an environment which is very hostile to technology. There are no precise figures on the globally available potential for energy production from tidal currents, but at least 1,200 TWh per year should be sustainably usable. This would correspond to over 6 % of the current worldwide electric power consumption [1].

Physical Principles

Tidal-stream turbines make use of the kinetic energy of the tidal flow. However, they cannot convert all of this energy to electricity, since the water must flow on past them. According to the so-called Betz's Law, the maximum power that can be extracted is given by:

$$P_{\max} = \frac{16}{27} \frac{\rho}{2} A \cdot v^3 \tag{1}$$

Here, ρ is the density of the fluid, v its flow velocity, and A the area of the turbine rotor. The available power thus increases as the cube of the flow velocity. Figure 1 makes this clear: it shows the maximum theoretically available power as a function of flow velocity for various rotor diameters. One can readily see that suitable sites should have flow velocities of at least 2 to 2.5 m/s; otherwise the power yield is too low.

Ideal locations have flow velocities of 4 m/s and higher. Such high velocity tidal streams are to be found for example in the Pentland Firth (Scotland), in the Bay of Fundy (Canada), or in the English Channel around the Island of Alderney.

The flow velocity changes with the tides and thus with time. Figure 2a shows typical measurements of the flow velocity over a month's time. One can discern the variations with the daily tides and also with the lunar cycle (spring tides and neap tides). Figure 2b gives the "tidal ellipse", a diagram which shows the direction and velocity of the flow (the points represent the arrowheads of the velocity vectors) during a tidal cycle. In contrast to wind energy, the tidal stream has a well-defined direction.

Challenges

Often, tidal-stream machines are compared with "underwater windmills". The developers of this technology must however deal with completely different challenges from those faced by wind-power plants. In particular, the turbines must be sealed against the entry of seawater. Tidal-stream turbines are thus more like a hybrid between a windmill and a submarine. The greatest challenges are:

- Accommodation to the periodically changing flow direction;
- Sealing of the rotor shaft under water;
- Corrosion protection;
- Environmental protection, if possible oil-free operation;
- Protection of the underwater electrical cables; and
- Minimal installation and maintenance costs. With a completely submerged construction, it must also be straightforward to locate the turbine for maintenance work.

While the direction of tidal flows is very predictable and the mechanical technology of the plants can therefore be kept fundamentally simple, the environment in which the machine must operate represents the real challenge. In order to keep operating costs under control, reliability and robust construction must have the highest priority in designing a machine which operates completely under seawater.

Case Study: The Technology of Voith Hydro

At *Voith Hydro* in Heidenheim, Germany, one of the world's leading hydroelectric equipment manufacturers, the Ocean Energies division is developing, testing and marketing tidal-stream installations. A prototype on a 1:3 scale with an output power of 110 kW and a rotor diameter of 5.3 m (see photo on p. 95) was installed in the winter of 2010/11 off the coast of South Korea's Jeollanam-do Province, near the Island of Jindo. Following this test run, beginning in 2015, a large tidal-stream park with an overall output of several hundred megawatts is to be constructed.

A tidal-stream plant can be subdivided into three main functional groups: foundation, turbine, and the installation and maintenance equipment. For suitable foundations, there are highly specialized firms (which also serve offshore wind technologies). *Voith Hydro* (VH) concentrates on the development of the mechanical and electrical components of a tidal-stream power plant, i.e. the turbogenerator, the land station, and the control technology. In addition, a concept for installation, retrieval, and maintenance of the turbine nacelle at moderate costs must be developed – and it must be usable even in remote locations with a poor infrastructure.

VH chose a completely submerged construction. Such plants are invisible from the coast and can be standardized to a great extent. Furthermore, the materials costs for the mounting towers are minimized. However, for maintenance the turbine must be retrieved and removed from the water.

In order to maximize the reliability and simplicity of the plant, VH turbine technology dispenses with complexity as far as possible. Important basic characteristics are:

- Torque control (variable speed) instead of a mechanism for changing the turbine-blade pitch with complex electronics (corresponding to pitch control for wind plants);
- Energy conversion in both flow directions by reversal of the rotation direction: symmetric rotor-blade profiles allow this, and are optimized for high efficiency and a wide application range;
- Direct drive, since a transmission can fail and requires lubrication with oil;
- Permanent-magnet excitation in the generator, to avoid the complexity of stator excitation and the use of slip rings;
- Direct cooling of the generator stator using ambient water from the tidal flow;
- Controlled water flow through the turbine to avoid rotating seals;

FIG. 2 | TIDAL FLOWS

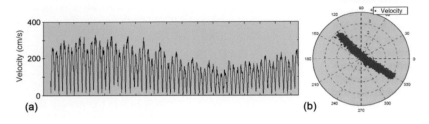

(a) (b)

(a) The measured flow velocity at a depth of 20 cm throughout one month (source: Renetec). (b) The tidal ellipse.

- Seawater-lubricated bearings, to eliminate the need for regular lubrication; and
- Oil-free design, to minimize dangers to the environment.

This system thus has only two moving parts: the turbine shaft (Figure 3) and the brake, which is needed for safety reasons. Everything else is either static or is operated passively. In the opinion of the authors, such a revolutionary design can provide the answer to all of the challenges of the tidal environment.

Installation, Retrieval and Maintenance

The installation, retrieval and maintenance of the plants should be as cost-effective as possible. If special ships or floating platforms are necessary, it must be possible to re-fit them in the simplest possible way. For the VH concept, a barge with a crane at its stern is required. The crane carries a framework, called the nacelle-retrieval module (NRM). It can be raised and lowered by chains and winches on the deck of the barge. Two guide chains are attached to the foundation of the plant.

Using a camera, or sonar if the water is cloudy, the NRM finds its way along the guide chains to the turbine (Figure 4). As soon as it has docked onto the turbine nacelle, it is locked in place by a hydraulic clamp. Then the NRM together with the turbine nacelle is raised to the surface by

FIG. 3 | TURBINE NACELLE

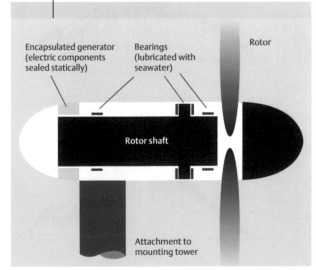

Encapsulated generator (electric components sealed statically)

Bearings (lubricated with seawater)

Rotor

Rotor shaft

Attachment to mounting tower

Seawater streams through the installation designed by Voith Hydro, and it has a minimum of moving parts (graphics: Voith Hydro).

Wave Power Plants

Energy Reserves from the Oceans

BY KAI-UWE GRAW

An old dream of humanity is to make use of the almost immeasurable energy of ocean waves. Their destructive power has however up to now not permitted any economically reasonable design to survive for long, although there have been many promising attempts and approaches.

The reduction in carbon dioxide emissions, which is increasingly urgent and is a major goal of governments and societies, has conferred a new significance on wave energy – as on all renewable energy sources. Interest in wave energy power plants, which could make appreciable contributions to the world's energy supply, is steadily growing. The use of wave energy for generating electric power has been under investigation for many decades. However, the countless, sometimes extremely naive suggestions for the application of wave energy have given this renewable energy source a dubious aftertaste in the public mind. But the long-term commtiment of a few research teams is now leading to a rethinking of this view.

The ocean waves contain inexhaustible reserves of energy. They are estimated to store around ten million terawatt hours of wave energy per year. This makes them, in principle, very attractive as an energy source. However, large waves can deploy a destructive force which makes huge demands on the stability of the wave power plant under load. It is thus particularly interesting to employ wave power plants precisely where the force of large waves must in any case be broken: along coastal protection installations. Conventional breakwaters only reflect or dissipate the wave energy without making use of it. Wave power plants, in contrast, extract the energy and convert it into useful electric power. Furthermore, the use of a breakwater as the structure for a power plant reduces the cost of the plant which is integrated into it. Figure 1 shows a new wave power plant in the Spanish harbor Mutriku, whose construction was supported by the EU.

Fig. 1 *The 300 kW breakwater power plant in the harbor of Mutriku, Spain, operates on the oscillating water column (OWC) principle. It went online in 2011* (photo: R. Wengenmayr).

With 16 Wells turbines, each delivering 18.5 kW, it can supply 250 households with electrical energy. Wave power plants which convert energy without fulfilling any protective function can even take the form of free-floating installations in the ocean (Figure 2).

Because of the stringent requirements for robust construction and corrosion resistance, the overall role of wave power plants is still very minor. Their currently-installed total output power worldwide is barely 3 MW; about five firms offer commercial installations, while another 20 or 30 are in the testing phase with their concepts [1].

The Formation and Propagation of "Gravity Waves"

The major portion of the energy which is stored in ocean waves is transported by so-called gravity waves (Figure 3). They are produced by the wind, and their motion is governed almost entirely by the gravitational force. Figure 3 also illustrates the forces which activate the waves: Short to medium waves are mainly produced by winds, longer ones by air-pressure differences due to weather fronts or by earthquakes; extremely long waves are due to the tidal forces. The figure additionally shows the three forces which in general govern the propagation of the waves: surface tension, gravity, and the Coriolis force. This last force is due to the earth's rotation; it is weak and has a noticeable influence on waves only when they are several kilometers long. The

INTERNET

LIMPET
www.wavegen.co.uk

OWC Wave Power Plants (animation)
bit.ly/PCzCWa

Wave Dragon (infos, docuvideo, animation)
www.wavedragon.net

Archimedes Waveswing
bit.ly/PXLN3v

Buoy System "WaveBob"
wavebob.com

surface tension of the water is also a very weak force. It is important only for waves that are shorter than about one centimeter: Such waves are flattened by the surface tension. In all other cases, gravity waves predominate. The force of gravity pulls the water in the wave crests down towards the troughs and thus tends to equalize the differences in height, acting as a restoring force. A simple introduction to the theory of waves is given in [2].

Water waves produced by the wind are generated mainly over deep water. Their form depends on the wind velocity, the duration of the wind, and the distance they have propagated since they were generated. The regions of the ocean surface with the most wave energy are therefore the open oceans, far from the equator (Figure 4). The wind is subject to friction with the surface of the ocean water; it pushes on individual water particles and thus accelerates the water layers near the surface. Turbulences in the airflow give rise to pressure differences between different parts of the water surface. To equalize these differences, the surface rises and sinks. This now-rough surface is subject to ever stronger pressure differences from the wind, which in turn increase the amplitude of the surface roughness. In this way, higher and higher quasi-periodic waves are formed.

Wave dynamics in the end limits further growth of the waves. The simple model of "linear wave theory" already gives a realistic value for the maximum wave height. It is ca. 14 % of the wavelength. According to linear wave theory, the individual water particles in the wave attain speeds at the tops of the wave crests which are greater than the propagation velocity of the waves themselves. They practically "fall out" of the wave in the direction in which it is moving. At this maximum height, the wave thus becomes unstable, and the wave crests form foamy whitecaps. Turbulence consumes part of the wave energy. When the waves have reached their maximum height and their period no longer changes, even if the wind continues to blow, the sea condition is called a "fully developed sea". After the wind has died down, the waves can maintain their energy

over distances of many thousands of kilometers. They are then referred to as "groundswell".

When the waves move into shallow water, their length and velocity decrease. Friction with the ocean bottom dissipates their energy and often changes their direction. When the wave velocity has dropped to a certain limiting value, the waves break and form a foamy surf. The breaking waves also lead to energy loss through turbulence. In planning coastal wave energy power plants, this process must be taken into account.

The Basic Technology for Exploiting Wave Energy

Beginning in 1986, a simply-constructed wave power plant was built in Norway, where it was operated for about twelve years. The plant was on the island of Toftestallen near Bergen, and it was intended as a demonstration project for interested groups. The TAPCHAN (TAPered CHANnel) directs the water from the incoming waves into a channel which rises and narrows, then empties into a raised basin. The water then flows steadily from this reservoir back to the ocean. In the process, it can power a conventional low-pressure turbine. The channel of the prototype plant had a 60 m wide opening on the incoming wave side and was between 6 und 7 m deep. The reservoir was at a height of 3 m above sea level. The incoming waves were steepened by the trumpet shape of the channel, so that they overflowed its banks, allowing the water to flow from the sides of the channel into the reservoir. The reservoir was lower than the end of the channel, preventing the water from flowing back out.

◄ **Fig. 2** *The WaveDragon is a floating ramp, which narrows down from a broad entrance and thus concentrates incoming waves. When the water is pushed up into a reservoir basin, it flows through a turbine. The first prototype, weighing 237 tons, is in Limfjord, Denmark, and delivers 20 kW of output power* (photo: www.wavedragon.net).

FIG. 3 | WAVES AND THEIR ENERGY

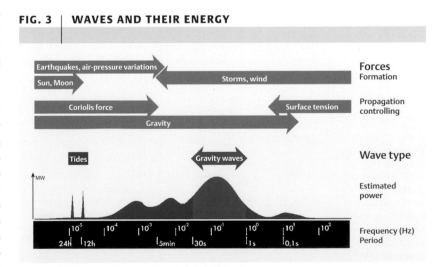

The power distribution of waves which act at the seacoasts, as a function of their oscillation period; gravity waves (red) contain the major portion of this power. Their period lies between one and thirty seconds. Above, the forces are indicated which predominate in the formation and propagation of the various types of waves.

FIG. 4 | DISTRIBUTION OF THE AVERAGE WAVE POWER

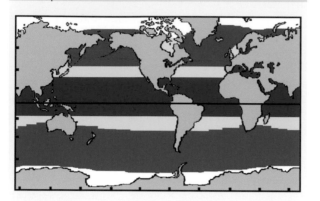

The distribution (simplified) of the average wave power density in the oceans; blue: 10–20 kW/m², yellow: 20–30 kW/m², red: 30–90 kW/m². This distribution corresponds in a good approximation to the distribution of wind energy.

The TAPCHAN in practice even exceeded by a small amount its planned output of 350 kW maximum power and 2 GWh annual energy production; this is so far a notable exception for wave energy power plants. Operating problems arose from earthslides after strong rainfall, pieces of rock flushed into the channel by the sea, and damage to the channel walls. In contrast to these constructional difficulties, the power generation with a standard water turbine caused no problems. However, there were no follow-up projects – probably because reserving large coastal regions for such storage power plants would not be economically feasible. Coastal regions are exploited today in so many diverse ways that the integration of new, area-intensive uses would hardly be enforceable.

In retrospect, one sees that the basic concept of this plant, i.e. providing electric power continuously and as required, is today relevant only for island systems. Large power grids, in contrast, can accept the fluctuating power generated by wave plants. Such systems are therefore currently being developed further with much smaller water storage basins, which do not "smooth out" the power output so effectively. Two examples of such projects are the Norwe-

gian Slot-Cone-Generator, a land-based system [3], and the European WindDragon, a floating system (see Fig. 2).

Today's Standard Technology: the OWC

OWC is the abbreviation for Oscillating Water Column and describes a design which makes use of the wave motion in rising and falling water columns. Figure 5a shows how a typical OWC functions. It consists of a chamber with two openings. One of these is on the side towards the incoming waves and lies beneath the water level. The water can enter the chamber through this opening, driven by the wave energy. The second opening allows the pressure to equalize with the surrounding air. The water column in the chamber moves up and down at the wave frequency, and thus "breathes" air in and out through the second opening. This "breath" drives an air turbine. It is constructed in such a way that it converts the oscillating motion of the air column into continuous rotational motion (see the infobox "Operating Principles of OWC Turbines" on p. 89).

In principle, an OWC represents a simple motion-conversion device (Figure 5b): In order to drive the generator, it converts the strong force at low velocity of the wave motion into a motion of the air column with a weak force but a high velocity. The essential aspect is that the low specific mass of the air permits a high acceleration.

OWC systems have been in use for decades for the energy supply of beacon buoys (Figure 6). They were invented by the Japanese Yoshio Masuda. In an OWC buoy, a vertical pipe assumes the function of the chamber. It reaches down into the calmer water layers below the buoy. Therefore, the water column in the pipe is at rest relative to the waves outside – but it moves relative to the buoy, since the latter is raised and lowered by the wave motion. Like standard OWCs, most buoys employ an air turbine. Such buoys have rather quickly become standard for applications that require limited power outputs. Some of them have survived more than twenty years of operation at sea.

FIG. 5 | THE OWC PRINCIPLE

The power plant converts the oscillating motion of the water column into elec-trical energy (a). It acts like a step-up gear box (b).

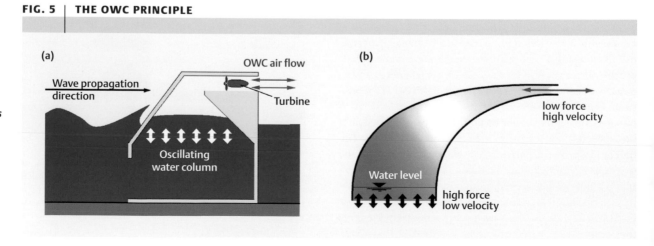

Existing OWC Projects

Between 1978 and 1986, in an international experiment initiated by Japan, an OWC system was tested for the first time on a large scale: The ship 'Kamai' carried out three series of tests in the Japan Sea in which turbines with up to a megawatt of power output were installed. Only in 1998 was the idea of a floating OWC again taken up in Japan. The new Japanese prototype 'Mighty Whale' however has a power output of only 110 kW (Figure 2). An Irish group is currently testing a reduced-scale floating construction at sea.

It has taken a similarly long time for the first continuation of coastal projects to be pursued, with an OWC of relatively high power output. From 1985 to 1988, the *Kværner* company in Toftestallen, Norway, tested an OWC built on the rocky coastline with a power output of 0.5 MW; it was constructed mainly of steel. Only since the end of the year 2000, on the island of Islay in Scotland, has a coastal OWC with a similar power output once again been in operation: LIMPET (Locally Installed Marine Power Energy Transformer, Figure 7). It was planned for an output power of 500 kW, but achieved only 250 kW in practice. It is constructed for the most part of concrete, as is a very similar project on the island of Pico, in the Azores, with 400 kW output power. The concrete design has been tested beginning already in 1983 in Sanze (Japan), from 1990 on the island of Dawanshan (China), and from 1988 on the island of Islay with considerably smaller prototype plants. Whether concrete will prove to be more enduring than steel is still under debate by the experts.

Breakwater structures usually consists of many concrete cubes of the size of a single or multiple-family house. For breakwater wave-energy power plants, one or more such cubes are modified in such a way that they can be employed as OWC chambers. The construction of these OWC chambers directly at the locations where they are to be used has proved to be the main problem for all of the test projects mentioned. At a location where waves are breaking onto the shore, people have to work continuously for several months – a dangerous, difficult and therefore expensive undertaking. For this reason, those construction engineers who built OWCs near Trivandrum (India) in 1990 and Sakata (Japan) in 1988 took a different route: Both were built as concrete caissons. These caissons were produced using established methods in a drydock. The firm *ART* (*Applied Research and Technology*) also built its steel wave-power plant OSREY (Ocean Swell powered Renewable EnergY) with 500 kW of output power at a shipyard. ART-OSREY demonstrated that even the installation of a previously-constructed wave breaker has its uncertainties: during the installation of the power plant in water 20 m deep off the Scottish coast, a severe storm came up. The structure was not designed to withstand such stresses during its installation phase, and it was destroyed.

All of the OWCs so far built have to be classed as test installations, with which construction techniques can be tried out and turbine technologies developed – although the Indian OWC has already fed power into the local grid. ART-OSREY was supposed to be the first commercially-operating OWC prototype. A current demonstration project has just been built in the Spanish harbor of Mutriku, as mentioned above (Fig. 1). The first completely commercial breakwater-OWC power plant, with an output of 4 MW, the Sidar Wave Energy Project (SWEP), is planned for the near future on the Scottish Hebrides island of Lewis [1].

Technological and Economic Questions

Why does wave power not yet come "out of the wall socket"? As we have already described, the installation of the power plants in the rough locations where they must be operated is difficult. Their design must guarantee a long operating life. A further hurdle is the developement of turbines which are suitable for OWCs. The turbines used up to now do not perform satisfactorily: Their efficiencies are too low and their constant velocity operation is problematic. Wells turbines so far achieve efficiencies only in the range of 50 to 70 %. Conventional turbines, in contrast, operate with up to 90 % efficiency. Even though they produce electrical power in only one flow direction, they can still be considered as serious competition for new designs.

OWC designs have also been tested which supply conventional turbines with a uniform air flow and thereby compensate their disadvantages. For example, the 30-kW Kujukuri OWC, which was built in 1987 in Japan in the Kujukuri harbor, uses pressure storage vessels for the air which is compressed by the waves. The storage vessels supply conventional turbogenerators without reversing the air flow.

FIG. 6 | AN OWC BUOY

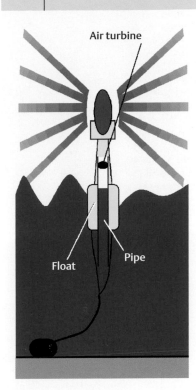

Air turbine

Float

Pipe

OWC buoys are for the most part designed as beacon buoys, with their own autonomous energy supplies.

Fig. 7 LIMPET is a coastal OWC with an output power of currently up to 250 kW, that has been in operation on the island of Islay (Scotland) since the year 2000.

Fig. 8 *The Archimedes Wave- swing consists of submerged cylinders which float in the water. The outer cylinder (green) moves up and down relative to the inner one (black). To generate power, linear generators (gray) utilize this motion – reversing the principle of the linear motors on a magnetic levitation railway.*

An additional technical problem is the quality of the electrical "wave power": as mentioned, it fluctuates, and the fluctuations must be compensated by the power grid. As in the case of wind power plants, the power production varies with changing weather conditions, and depending on the location of the plant, the tidal variations add to the fluctuations. For OWC power plants, there is in addition a periodic fluctuation which reflects the relatively high frequency of the incoming waves and is passed on to the power grid.

To be economically feasible, power plants must be planned for an operating lifetime of at least twenty years, while their "moving parts" should last at least ten years. In estimating the financial boundary conditions for the use of wave energy, a fundamental physical property of the waves must be considered: Their energy increases as the square of their amplitude or height. To illustrate the economic and technical restrictions, let us consider an example: Suppose that a wave energy converter is designed to extract energy from waves that are one meter high. In order to withstand extreme storms, however, at the same time it has to be able to deal with waves that are roughly ten

times higher – that is, waves ten meters high. Such waves carry wave energies which are a hundred times greater than that of the waves for which the plant is designed! This requirement can cause the construction costs to explode in comparison to those of other types of power plants.

Other Technologies for the Exploitation of Wave Energy

As mentioned at the beginning, there are numerous different ideas for extracting energy from ocean waves. Two of them are distinguished by their special designs. Their developers are trying to take the bull by the horns and solve the main economic problem of wave energy plants: the requirement that the plant be able to survive "monster waves". Both systems have been built as prototypes.

Figure 8 shows the Archimedes Waveswing. It consists of submerged cylinders which are anchored to the ocean floor; they are 21 m high and 10 m in diameter. A wave which passes above this structure gives rise to a "dynamic" variation in its buoyancy; the oscillating water flow moves the closed cylinders up and down. This design has the advantage that it is not subject to the strong wave forces near the surface of the water. Its disadvantage: in the deeper water levels, only a small fraction of the wave energy is still available. If the cylinders were anchored near the surface in shallow water, then the rolling motion of the waves would cause strong horizontal stresses; the cylinders cannot convert these motions into usable energy, they only add to the load on the structure. The first prototype of the Archimedes Waveswing was installed at a site off the Portuguese Atlantic coast, and fed electric power of up to 1.5 MW into the Portuguese power grid.

The builders of Pelamis (Figure 9) asked themselves the question as to how an energy converter would have to be designed in order to survive large waves with the lowest

Fig. 9 *The prototype of a Pelamis plant, which was installed in 2004 off the Scottish Orkney Islands* (photo: Ocean Power Delivery Ltd.).

construction costs. The result is a system which can be strongly deformed by smaller waves, and converts their energy in an optimal fashion, but will still not be damaged by large waves. Since Pelamis is a kind of "snake" made of movable, coupled segments, it was given the name of a genus of sea snakes. With a total output power of 750 kW, the prototype Pelamis is around 150 m long and 3.5 m in diameter.

Normal sea conditions, with a relatively moderate vertical oscillation, cause the segments of Pelamis to perform a horizontal evasive motion, like a sea snake swimming. Hydraulic assemblies convert this motion into usable energy. They produce a high pressure whose energy is stored in intermediate air pressure chambers. These chambers then feed hydraulic turbines which operate at constant speed. Pelamis thus converts brief wave impulses with a high power into a constant, lower power output from the turbines. For this reason, the pumping power capacity of the hydraulic pumps must be designed to be considerably greater than the power output of the hydraulic turbines.

When confronted by large waves, on the other hand, Pelamis acts like a stiff structure. It cannot follow the large vertical oscillations, and simply dives through the waves. Since its structure has a relatively small cross-sectional area, it is subject to only moderate forces when diving through large waves. For that reason, this new approach is very promising. After the prototype was tested for several years off the Scottish West Coast, an improved sea snake was constructed: Pelamis P2 is 180 m long, 4 m in diameter, and weighs in at 1300 tons [5].

Can Wave Energy Soon Be Commercially Exploited?

The energy crisis in the 1970's aroused strong interest in renewable energy sources. Thus, the use of wave energy for electric power generation, like many other renewable energy sources, became the subject of intense research. After the end of the crisis, the technology was however deemed too expensive and was put aside. Only a few, for the most part Asian research institutions continued working on the topic. In Europe, this situation changed only in the 1990's. At that time, the European Union included wave-energy conversion in the research program JOULE. A major part of the European prototypes described here are based upon this research effort. In Germany, there is no official program of research support for wave energy. In Denmark, in contrast, a broad-based program is in place, which is intended to repeat the successes achieved in the use of wind energy.

This European research is today most certainly the motor for further developments. In Asia, however, the research projects have thus far not been successful. The Japanese energy providers do not want to feed the fluctuating power from wave-energy plants into their grid. In India, it was expected that energy production costs should be at the level of conventional (fossil-fuel) plants. Both of these demands

---

OPERATING PRINCIPLES OF OWC TURBINES

In wave power plants, the turbines have to withstand about twenty load alterations every minute. In OWC power plants, these vary not only between zero and maximum flow rates, but they also periodically reverse the direction of flow. The development of frequency transverters for wind turbines has made it possible that the generators no longer require a constant rotation speed. But so far, the direction of rotation must be kept constant. In order to solve this problem, there are two approaches in use today: the Wells turbine and the impulse turbine. Both types are currently being tested to determine whether they are serviceable for wave power plants.

Wells Turbines

The Wells turbine has blades with a symmetric profile which is perpendicular to the air flow. Once they are set in motion, they maintain their direction of rotation, even if the direction of the air flow changes. When the turbine is rotating, the overall approach flow to the blades is composed of two components: one component is the air which is flowing into or out of the OWC, the other is the approach velocity of the airfoil, which depends on its rotational speed. The two components add to give an overall velocity. It makes an angle with the airfoil which depends on the two flow velocities. The resulting reaction force on the airfoil is perpendicular to the overall flow. The reaction force can be further decomposed into a driving force and a buoyancy force.

The buoyancy force simply pushes against the turbine bearings and is not useful; only the driving force contributes to the rotation of the turbine. If the direction of the OWC air flow reverses, the force diagram remains the same. This decomposition of forces makes it clear why a Wells turbine has a lower efficiency than a conventional turbine.

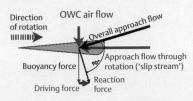

Distribution of forces on one blade of a Wells turbine.

Impulse Turbines

Those designs, which redirect the approach flow within the turbine rotor, are much simpler. Impulse turbines have fixed guide vanes for this purpose (shown black in the figure), between which the turbine rotor (blue) revolves. The figure shows (as dashed red lines) how the guide vanes direct the OWC air flow in such a way that it produces a driving force on the blades of the turbine rotor. Reversal of the flow direction does not change the direction of rotation of the rotor. The impulse turbine has the disadvantage that losses occur in the redirection of flow before and after the turbine rotor, as well as in the gaps between the turbine blades and the guide vanes. For this reason, it initially could not compete successfully with the Wells turbine. Its superior constant-velocity properties could however soon put it back on the map.

Operating principle of the Wells turbine.

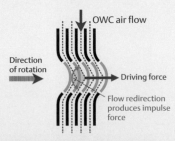

Operating principle of the impulse turbine.

were discussed in Europe at the time when wind energy was being introduced, and have to a large extent been put into perspective. For example, in practice it has become apparent that the power grid can buffer the fluctuations due to wind power feed-in without problems. It would also be helpful for the comparison with conventional power generation if all external costs (environmental effects, resource consumption, risks) were to be included in the computations. This has for the first time created economically acceptable boundary conditions for the commercial wave-energy projects. Developments such as the entry of the large water turbine manufacturer *Voith Hydro* in Heidenheim, Germany into the area of sustainable energy, through its purchase of the Scottish pioneer *Wavegen*, show that wave-energy power plants are now on the threshhold of commercial success.

How Expensive Would Wave Energy Be?

A rough estimate of the most important cost factors yields the following results: The price per kilowatt hour from installations which are integrated into breakwaters and can therefore be favourably costed, would, according to the estimate of *Voith Hydro*, be 15 €-cent/kWh [6]. This is already notably lower than the cost of photovoltaic power in the foreseeable future (see the corresponding chapters in this book).

The energy price for floating systems is in fact strongly dependent on the costs of underwater power cables, which are required to transport the current generated. The increasing number of offshore wind-energy parks will certainly lower the production costs of these cables. Furthermore, combined wave and wind energy power plants could utilize the same underwater cables cooperatively. This would effectively halve the cable costs.

An additional argument in favor of the use of ocean wave energy is the European electric power consumption in practice: In winter, power consumption increases. At the same time, the North Atlantic weather situation produces more wave and wind energy on the European seacoasts. In comparison to solar energy, which is mainly available in the summer, these two energy sources thus conform much more closely to the seasonal energy consumption patterns in Europe.

Conclusions

The technological and commercial boundary conditions today – for the first time – are bringing the use of wave energy into reach. For European energy policy, it will be decisive whether the current prototype plants operate reliably in the coming five to ten years. Success here would stimulate the construction of commercial wave power plants. With an increasing number of plants, the technology could be optimized and would certainly become less expensive. Then, the immeasurable energy of the oceans could make a perceptible contribution to the energy supply for humanity on a long-term basis.

Summary

Prototypes of wave power plants have been tested for several decades, mostly utilizing the technology of an oscillating water column (OWC). Autonomous OWC lighthouse buoys have already successfully established themselves. Larger power plants have been tested in the form of floating and stationary prototypes with power outputs of up to a megawatt. The main problem and most serious cost factor are extreme waves, which can destroy such systems during installation and operation. The European Union is now supporting pilot projects with the goal of introducing commercial wave plants if the pilot plants are successful. Electrical power from wave plants would presumably cost about the same as wind-generated power.

References

[1] J. Weilepp, WasserWirtschaft **2009**, *3*, 18.
[2] H. Vogel, Gerthsen *Physik*, 18th edition (Springer-Verlag, Heidelberg **1995**), p. 197.
[3] waveenergy.no, see under "Technology and Innovation".
[4] www.awsocean.com, see under "Technology".
[5] www.pelamiswave.com/our-technology/the-p2-pelamis.
[6] R. Wengenmayr, Physik in unserer Zeit **2008**, *39 (1)*, 6.

About the Author

Kai-Uwe Graw, born in 1957 in Berlin, obtained his doctorate in Structural Engineering in 1982 and his doctorate in 1988, both at the Technical University of Berlin. He received his postloctoral lecturing qualification for civil engineering in 1995 at the Bergische Universitaet – Gesamthochschule Wuppertal. From 1996, he was professor for substructural and hydraulic engineering at the University of Leipzig, and since 2006, at the Technical University of Dresden.

Address:
Prof. Dr.-Ing. habil. Kai-Uwe Graw, Institut für Wasserbau und Technische Hydromechanik, TU Dresden, 01062 Dresden, Germany.
kai-uwe-graw@tu-dresden.de

Osmosis Power Plants
Salty vs. Fresh Water

BY KLAUS-VIKTOR PEINEMANN

One possibility of obtaining sustainable energy from seawater is the use of osmosis. The key to this technology is the development of efficient membranes which allow water to pass through, but not salt.

Osmosis is an omnipresent process, which for example causes sausages to burst in hot water. Osmosis plays an important role in every living cell for materials transport, and it allows trees to pump water up to great heights. It can also be used as a source of sustainable energy everywhere where fresh water flows into salty seawater at the mouths of rivers.

The Norwegian energy concern *Statkraft*, which has been working on the development of such osmosis power plants since 1997, estimates the global potential of this technology to be in the range of 1600–1700 TWh of electrical energy per year [1]. For comparison: In the year 2008, all the hydroelectric plants on earth produced about 3200 TWh, and the nuclear plants produced over 3100 TWh [2]. For Europe, *Statkraft* estimates a potential of 200 TWh/a, and for Norway, 12 TWh/a [3]. There, in the town of Tofte on the Oslofjord, *Statkraft* put the world's first osmosis pilot plant into operation in the fall of 2009.

Osmosis occurs whenever a semipermeable membrane separates solutions of a differing salt concentration (or sugar concentration, etc.). 'Semipermeable' means that the membrane allows water to pass through, but not the dissolved salt. If such a membrane is between a salt solution and pure water, the water tends to pass through it into the salt solution. The reason for this is that the fraction of pure water in the salt solution is lower, so that the flow of water through the membrane tends to equalize this concentration difference (diffusion).

On the saltwater side of the membrane, the pressure is thus increased, depending on the concentration of the salt solution. This additional pressure is called 'osmotic pressure'. With a suitable installation, it can be utilized to drive a turbine and generate electrical power (Figure 1). Some portion of the diluted saltwater must however be diverted off the flow before it reaches the turbine, in order to pressurize the newly-arriving saltwater to the level of the osmotic pressure. In contrast to the simplified schematic in Fig. 1, the thin membranes are in fact rolled up into compact modules.

The osmotic pressure can reach rather high values. In blood, it is for example 7.5 bar, while for seawater, with a salt concentration of 33g/l, it is 25 bar. This means that a column of water 250 m high could be supported on the seawater side if seawater and fresh water are separated by a semipermeable membrane. However, only about half of the theoretical maximum pressure can be utilized technically, about 13 bar – which still corresponds to a water column 130 meters in height. This would be a very effective water head for hydroelectric plants.

The idea of constructing osmosis power plants at locations where rivers flow into the sea came into play rather early. In 1974, two patents on energy production through osmosis were independently applied for in the USA alone. The pioneer of osmosis power plants is considered to be the Israeli Sidney Loeb, who introduced his concept in 1975 [4].

For a long time, there was little interest in this idea. The available membranes at the time were much too ineffective, and there was no hope of being able to compete success-

A schematic of an osmosis power plant. Salty and fresh water flows through a basin with a semipermeable membrane. Osmosis has the effect of pushing the fresh river water into the salty seawater compartment, whereby an overpressure results. A portion of the surplus water can flow out of this compartment and drives a turbine.

FIG. 1 | AN OSMOSIS POWER PLANT

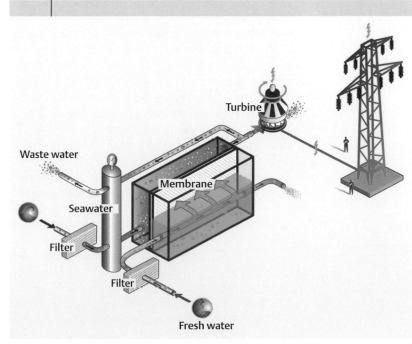

Turbine

Waste water

Membrane

Seawater

Filter

Filter

Fresh water

Fig. 2 *A scanning electron microscope image of an osmosis membrane. In the cross section, one can discern the support layer below and the filter layer above it.*

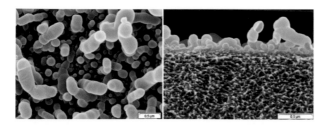

fully with the low power costs from fossil fuel plants. In the 1990's, engineers at *Statkraft* and the Norwegian research institute SINTEF (in Trondheim and Oslo) took up the idea again. They began research to find more effective membranes, together with several European partners. One of these partners was the Institute for Polymer Research at the GKSS Research Center in Geesthacht, Germany, where our group prepared membranes on a prototype scale for testing.

The starting point for this development were the membranes used for the desalination of seawater. This process is called reverse osmosis, since here, an external pressure which is notably higher than the osmotic pressure must be applied to the seawater side, causing pure water to pass through the membrane. By modifications of the membrane structure, it proved possible to increase the osmotic power of the membranes by a factor of 20. Today, at the King Abdullah University of Science and Technology in Saudi Arabia, we are working on the continued improvement of osmotic membranes.

The most efficient desalination membranes are composite materials. The actual desalination effect occurs within a layer which is only about $1/10,000^{th}$ of a millimeter thick. It is supported by a porous substrate of 0.1 to 0.2 mm thickness (Figure 2). Research and development of more effective osmosis membranes is concentrating particularly on this substrate layer, which must be as open as possible.

In addition to membranes derived from reverse osmosis technology, research groups around the world are working on completely new structures, which are as yet still dreams for the future. One idea is to use substrate layers made of parallel-oriented carbon nanotubes, which serve as tiny transport channels for the water. Scientists at the Lawrence Livermore Laboratory in California were able to demonstrate this in principle in 2006 [5].

The critical factor for success is the electric power which can be generated per square meter of membrane area. According to the calculations of the *Statkraft* engineers, an output power of 4 to 6 W/m² of membrane area is necessary in order to be competitive with other sources of sustainable energy. We have not yet quite reached this goal, but have come very close due to the development work in recent years. Our group achieved about 3 W/m² in 2009.

In addition, the newly-developed membranes must be feasible to manufacture in large quantities at a price of about 10 €/m² in order to make osmosis power competitive. Just to utilize 10 % of the estimated European potential, we

would require 700 million square meters of membrane material [1]. In considering this number, however, one must not forget that for desalination, already many millions of square meters have been manufactured.

The potential is great. Osmosis power plants can be built anywhere where fresh water flows into the sea and where the gradient of salt concentration is sufficiently large. The thermodynamically-possible maximum energy (reversible work) which is released when 1 m³ of fresh water comes into contact with seawater is 2.2 MWh [6]. Of course, in practice only a fraction of this energy can be used. In contrast to wind energy and solar power, an osmosis power plant is independent of the weather, and is therefore a genuine base-load power source. Among renewable energy sources, only geothermal energy and the combustion of renewable biomass are in this same category.

The challenges for the developers of membranes and plants are great. But the chance of being able to revive a thirty-year-old idea and make it into a new source of renewable energy spurs all those involved on to increased efforts to succeed with this exciting project.

Summary

A possibility of obtaining sustainable energy from the oceans is offered by the use of osmosis. Its worldwide potential has been estimated by the Norwegian energy concern Statkraft to be around 1600 TWh per year. This corresponds to 50 % of the worldwide power output of hydroelectric plants (as of 2008). Statkraft put the world's first prototype of an osmosis power plant into operation in the fall on 2009. It is located in Tofte, on the Oslofjord, Norway.

References

[1] www.statkraft.com/energy-sources/osmotic-power.
[2] World Energy Statistics **2010**, International Energy Agency, IEA/OECD, Paris 2010, p. 24.
[3] S.E. Skilhagen *et al.*, Desalination **2008**, *220*, 476.
[4] S. Loeb, Science **1975**, *189*, 654.
[5] R.F. Service, Science **2006**, *313*, 1088.
[6] K.-V. Peinemann *et al.*, Membranes for Power Generation by Pressure Retarded Osmosis, in: K.-V. Peinemann, S. Pereira Nunes (Eds.), *Membranes for Energy Conversion*, Wiley VCH, Weinheim **2007**.

About the Author

Klaus-Viktor Peinemann studied chemistry at the Christian-Albrechts University in Kiel and obtained his doctorate there (neues Photo) in 1982. Up to 2002, he directed the Membrane Development Division at GKSS; thereafter he served as Senior Scientist. In 2004, he was named Honorary Professor at the Leibniz University in Hanover. Since October 2009, he has carried out research at the King Abdullah University of Science and Technology (KAUST) in Saudi Arabia.

Address:
Prof. Dr. Klaus-Viktor Peinemann, Research Center, King Abdullah University of Science and Technology (KAUST), Thuwal 23955-6900 Saudi Arabia. klausviktor.Peinemann@kaust.edu.sa

DLR Studies on the Desertec Project
Power from the Desert

BY FRANZ TRIEB

By the year 2050, according to the DESERTEC concept, sustainable energy from the sunny regions of Southern Europe, North Africa and the Middle East could yield so much power that all these countries could cover their own needs and also supply 15 to 20 % of the power consumed in Northern Europe. This project is based on studies by the German Aerospace Center (DLR).

In the year 2003, the Club of Rome, the Climate Protection Fund in Hamburg, and the Jordanian National Energy Research Centre together developed the DESERTEC concept. The Desertec Foundation which was subsequently established, together with twelve industrial firms, published a Memorandum of Understanding in July, 2009, calling for construction of the project. On October 30, 2009, the industrial initiative *Dii GmbH* was founded; it is charged with preparing a business plan for practicable investments by 2012. The goal of these activities is to construct power plants in the countries of Southern Europe, North Africa and the Middle East by 2050 that can meet the energy needs of those countries from sustainable sources, mainly solar and wind energy, and in addition can supply 15 to 20 % of the remaining European power consumption (Figure 1).

The DESERTEC plans are based on studies in which the German Aerospace Center (DLR) determined the potential of sustainable energy sources for producing electric pow-

er and potable water in 50 countries in Europe, the Middle East and North Africa (abbreviated EU-MENA countries). The studies showed that sustainable sources in these regions could provide enough electric power and desalinated water to meet the needs of the Mediterranean countries and also deliver power to the rest of Europe. To generate this exported power, a land area of 2500 km² would be required, assuming that concentrating solar-thermal power plants (CSP) are utilised; this is roughly the area of the German state of Saarland. An additional 3600 km² would be needed for building high-voltage transmission lines. In this chapter, I summarize the results of our DLR studies, treating the aspect of water desalination only briefly.

The Increasing Demand for Electric Power and Water

As a first step for our analysis, we determined the demand for electric power and its projected evolution up to the middle of the 21st century. A part of the energy problem is also the growing deficit in the supply of potable water in the Middle East and North Africa (MENA), meaning that the demand for energy-intensive desalination of seawater will increase. For simplicity, we assumed that the energy required for this desalination, for example by reverse osmosis (see preceding chapter), would also be provided entirely in the form of electric power in the long term.

According to estimates by the United Nations, the population of Europe, now around 600 million, will remain stable, while the MENA region, which had a population of 300 million in the year 2000, will increase also to 600 million by 2050. Economic growth has two contradictory effects on the consumption of power and water: On the one hand, the demand for both increases in a growing economy; on the other, the efficiency of production, distribution and consumption also improves. In the past decades, a decoupling of economic growth and energy consumption could be observed in all the industrialized countries. In order to be able to afford measures to increase efficiency, a certain economic level above the bare existence minimum must first be attained. This precondition in the meantime holds in most of the MENA countries [1].

Our analysis shows that by the year 2050, the demand for electric power in the Middle East and North Africa will be around 3000 TWh/a (Figure 2). It will thus be comparable to the present consumption level in Europe. At the same time, it must be presumed that European consump-

A sketch of the possible infrastructure for delivering power to Europe, the Near East and North Africa (EU-MENA). CSP: Concentrated Solar Power, i.e. solar-thermal power; PV: photovoltaics (source: TREC/CoR).

Legend:
- Solar (CSP)
- Solar (PV)
- Wind
- Water
- Biomass
- Geothermal

Renewable Energy. Edited by R. Wengenmayr, Th. Bührke. Copyright © 2013 WILEY-VCH Verlag GmbH & Co. KGaA, Weinheim

tion will continue to rise and will stabilize near a value of roughly 4000 TWh/a (Figure 3). Due to increases in efficiency, our model yields lower values for the predicted requirements than some other scenarios [2]. On the other hand, there are also scenarios which are based on a lower demand [3,4]. A reduction of the demand in Europe after 2040 is indeed possible, but uncertain. A stagnation or slightly increasing demand is equally possible, since it may prove to be the case that new, energy-intensive services such as electric vehicles or hydrogen fuels for the transport sector will be required. We have not taken such possible paradigm changes into account in our study, but rather considered only the classical electric power sector.

We carried out a similar analysis for the water sector in the MENA countries. It clearly showed that these countries will be facing very serious problems in terms of their water supply in the not-too-distant future, if they do not soon take the necessary additional measures. Desalination of seawater is one of these additional options. Under the assumption that on the average, 3.5 kWh of electric power is required for the desalination of one cubic meter of water, this would lead to an additional requirement for nearly 550 TWh/a by 2050 for desalination alone. This corresponds to the current power consumption of a country like Germany, and must be added to the consumption values shown in Fig. 2.

Available Resources and Technologies

At present, the power consumers in the EU-MENA countries have no other choice but to pay the steadily-increasing price of power from fossil and nuclear fuels. This situation is exacerbated by the fact that fossil and nuclear fuels technologies today still receive ca. 75 % of all public subsidies for the energy sector. This figure would increase to over 90 % if external costs were also counted as hidden subsidies.

On the other hand, a number of sustainable-energy technologies are available (see the downloadable table at **www.phiuz.de** special features/Zusatzmaterial zu den Heften). Some of them produce energy on a fluctuating basis, for example wind energy and photovoltaic installations, while others could provide both peak-load and base-load power as needed. Among these are the biomass, hydroelectric power and concentrating solar thermal power plants.

The long-term economic potential of renewable energies within the EU-MENA region is much greater than its current power consumption; in particular, solar energy literally outshines all the others. The average energy of the annual insolation in the MENA region is 2400 kWh/m^2. If this energy were captured by solar-thermal power plants, up to 250 GWh of electrical energy could be obtained per year from a square kilometer. This is 250 times as much as from the biomass and five times more than with the current best wind or hydroelectric plants per square kilometer. A field of concentrating solar collectors on the same land area as the Nassar Lake in Egypt –more than double the size of the

A technician working on a parabolic-trough mirror at the Andalusian solar power plant Andasol. The central receiver tube is readily seen; here, the concentrated sunlight heats a thermal transfer fluid, which then drives a turbine and its associated electric generator (photo: Solar Millenium).

Saarland – would be able to generate as much energy as all the oil produced in the Middle East. In contrast, the Nassar Lake, which causes enormous ecological problems, yields only a fraction of Egypt's power needs.

In addition, there are also other sustainable energy sources in the EU-MENA region. There is a potential of 2000 TWh/a from wind energy, and an additional 4000 TWh/a from geothermal energy sources, hydroelectric power and the biomass. The latter includes waste products of agriculture and forests, as well as refuse from cities and sewage. Photovoltaics, wave and tidal energy also have a considerable potential in the region; to be sure, each of these sustainable energy sources has its specific geographic distribution (Figure 4), so that every sub-region exhibits its individual mixture of resources. Hydroelectric power, biomass and wind energy are the favored sources in the North, while solar and wind energy are the strongest sources in the South of the EU-MENA region.

Fossil energy sources such as coal, petroleum and natural gas represent a useful complement to the mixture of sustainable energies, since they can readily be used for en-

INTERNET

Background information on the DLR studies
www.dlr.de/dlr/en/desktopdefault.aspx/tabid-10200

The ecological balance for transmitting power from North Africa to Europe
www.dlr.de/tt/trans-csp

Desertec Foundation
www.desertec.org

FIG. 2 | **POWER CONSUMPTION IN MENA**

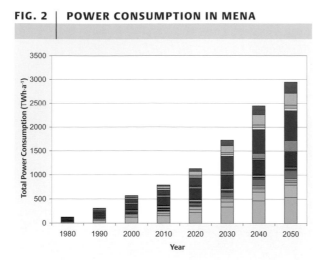

A scenario describing the power requirements of the MENA countries investigated: ■ *Egypt,* ■ *Saudi Arabia,* ◻ *Iran.*

FIG. 3 | **POWER CONSUMPTION IN EUROPE**

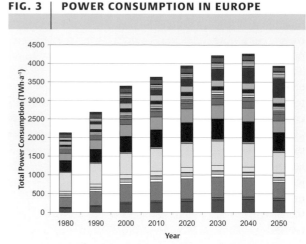

A scenario describing the power requirements of the European countries investigated: ■ *Turkey,* ▨ *Italy,* ■ *Great Britain,* ◻ *Germany,* ■ *France.*

ergy balancing and for securing the stability of the power grid. If their usage can be reduced to the point where they serve only to provide a reserve capacity, the current price rises will presumably be stopped and only a minor stress on the economic development will result. Furthermore, their environmental effects will be minimized. In addition,

their availability will be extended over a period of decades, if not centuries.

Nuclear power plants, in contrast, are less suitable for use in combination with sustainable energy sources, since their power output cannot be adjusted to a fluctuating demand, owing to economic constraints. Furthermore, the

FIG. 4 | **SUSTAINABLE ENERGY SOURCES FOR EU-MENA COUNTRIES**

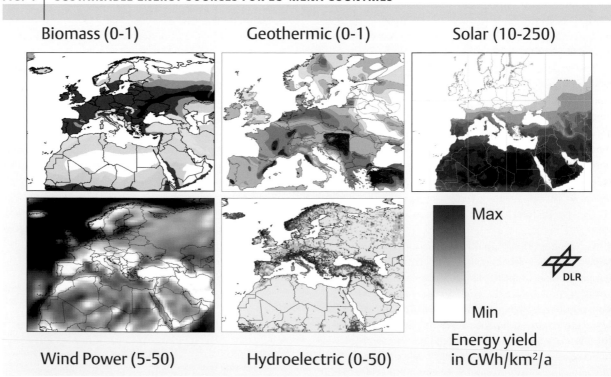

A map showing sustainable energy sources for the EU-MENA region, with the annual minimum and maximum electrical energy yields (in parentheses) that can be obtained from 1 km² of land area in each case. Solar energy includes both photovoltaics and solar thermal power. The darker areas are the most productive.

costs of decommissioning nuclear power plants exceed those of their construction, and the well-known problems remain, such as the uncontrollable proliferation of plutonium and the final disposal of nuclear waste.

Some of the sustainable energy sources are likewise able to supply power on demand for peak loads as well as for the base load, as needed. Among these are geothermal energy, hydroelectric plants in Norway, Iceland and in the Alps, as well as most biogenic sources and concentrating solar thermal power plants in the MENA region. These last, as solar-driven steam power plants, make use of the strong daily insolation in this area, which is relatively evenly spread over the year. And they offer the possibility of storing solar thermal energy for nighttime operation, as well as the option of using additional heat input from combustion of fossil fuels or biomass.

Solar-Thermal Power as a Key Element

Steam and gas turbines powered by coal, uranium, oil and natural gas are today's guarantees of power-grid stability. They provide base-load and peak-load power. However, turbines can also be operated with high-temperature heat from concentrating solar collector fields (Figure 5). Solar thermal power plants of this type with 30 to 80 MW output power have already been operating successfully in California for over 20 years, and new plants are being constructed right now in the USA, Spain and other countries, with up to 1000 MW of output power. By 2015, as much as 10 GW of solar-thermal power capacity could be installed worldwide, and by 2025, even 60 to 100 GW. At present, nearly 1 GW is online. According to a current study, today's solar energy price of 27 €-ct./kWh in Spain could decrease to below 10 €-ct./kWh [5].

The example of Andasol in Andalusia can be used to illustrate the principle (see the picture on p. 121). The plant consists of three installations of the same size, each using long, trough-like mirrors with a parabolic cross-section, similar to oversized rain gutters. In Andasol 1, an overall area of nearly two square kilometers contains more than 600 of these collectors, each one 150 meters long and 5.7 meters wide. All together, the mirrors have a surface area of more than 500,000 square meters (Figure 5).

Electric motors turn the collectors around their long axes to follow the path of the sun. The solar radiation, which falls perpendicular onto the collectors, is concentrated by a factor of ca. 80 in their focal line. Along this line, the absorber tubes are mounted. Their surfaces are covered with a special coating which absorbs sunlight especially well and converts it to heat. The tubes are filled with a synthetic oil, which is heated to nearly 400° C. The hot oil flows into a heat exchanger where its heat vaporizes water. As in a conventional power plant, the hot water vapor – steam – shoots into the turbines, which are connected to generators that produce electric power. The temperature of the steam at the turbine entrance is comparable to that in nuclear power plants. As in any steam power plant, the

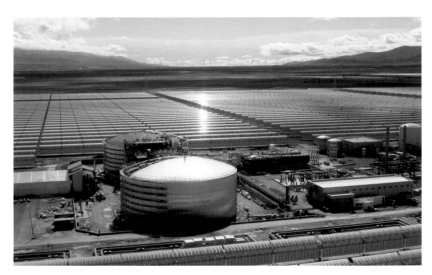

Fig. 5 *Andasol 1 and 2 with a thermal-storage reservoir in the foreground* (photo: Solar Millenium).

steam must be condensed when it exits the turbines. Since there is no cooling water in dry desert regions, dry coolers and so-called Heller cooling towers are employed there. These have proven to work well with conventional steam plants, for example at oil-fired plants in Saudi Arabia.

Experiments are also being carried out employing other heat-transfer media in the tubes. Molten salts can be used instead of oil; they can be heated up to 550° C and thus yield a higher efficiency for the turbines. Furthermore, it is possible to vaporize water directly in the tubes, thus avoiding the need for a heat exchanger. This technology is planned for a part of Andasol 3.

In contrast to solar cells, where the solar energy is converted directly into electric power, solar-thermal plants pass through an intermediate stage using heat energy. It can be stored on a large scale relatively easily in comparison to electrical energy, and this offers a considerable advantage. At Andasol, two large tanks of 14 meters height and 36 meters in diameter are used as heat-storage reservoirs. They are filled with a molten salt which is heated by the absorber fluid. Using this stored thermal energy, the plant can then deliver its full power for up to eight hours after sunset. Therefore, these plants are suitable for supplying both base-load and peak-load power, and also for providing balancing power to stabilize grid fluctuations. With an output power of 50 MW, Andasol 1 operates with solar heat input for 85 % of its annual power production; the rest is obtained from natural gas combustion.

It is also advantageous that solar-thermal plants can produce both heat and electric power. They can thus deliver steam also for absorption cooling machinery, industrial process heat, or thermal desalination of seawater.

Technological and Economic Questions

Taking into account the technological, social and economic boundary conditions, we developed a scenario for energy production in the 50 MENA countries up to the year

FIG. 6 | POWER GENERATION IN THE MENA COUNTRIES

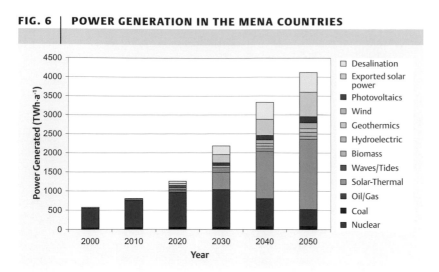

Power generated on the basis of sustainable and fossil energy sources in the MENA countries to supply the increasing energy demand. Exports of solar power to Europe and the additional power for desalination of seawater in the region are included.

FIG. 7 | POWER GENERATION IN EUROPE

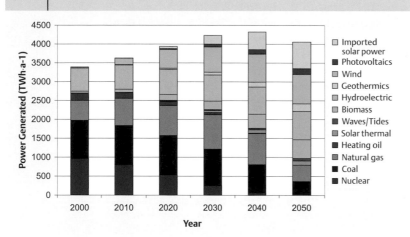

Power generated on the basis of sustainable, fossil and nuclear energy sources in European countries to supply the energy demand, including imports of solar power from the MENA countries.

2050. With the exception of wind and hydroelectric power, sustainable energy forms will hardly play a role in electric power generation there before 2020 (see Figures 6 and 7). At the same time, we assumed in our study that the phasing-out of nuclear power in many European countries and the stagnation of the combustion of lignite and anthracite coal for environmental reasons will lead to an increased utilization of natural gas. We used the official scenario of the European Commission up to 2020, which presumes a decrease in nuclear power plants of one percent per year.

Up to 2020, the growing fraction of sustainable energies will for the most part serve to reduce the combustion of fossil fuels. They can however also to some extent replace the

existing generating capacity for balancing power. Owing to the generally increasing demand for power and the replacement of nuclear plants, the combustion of fossil fuels will probably not be markedly reduced before 2020. The combustion of oil for power generation will have almost completely vanished by 2030 for cost reasons, followed by nuclear power in the last decade of the scenario. The latter will no longer be needed and will no longer be operable on an economically feasible basis due to the limited capacity utilization of conventional power plants. The consumption of natural gas and anthracite coal will increase in the intermediate period up to 2030 and then be reduced to a compatible and financially acceptable level by 2050. In the long term, it cannot be excluded that new types of consumers such as electric automobiles will cause the demand for power to again grow, so that an increased exploitation of sustainable energies will be necessary. Sufficient potential for this is in any case assured.

The mixture of electric power supplies in the year 2000 contained five sources, of which most are limited, while the projected mixture in 2050 includes ten sources, of which the majority are sustainable. For this reason, our scenario fulfills the declared goal of the European Commission in its "European Strategy for Sustainable, Competitive and Secure Energy", aiming at a greater diversification and security for the European energy base.

An essential condition for choosing a sustainable mixture of energy sources is the provision of secure power generation as needed, with a reserve of the order of 25 % in addition to the expected peak demand (Figure 8). Prior to the beginning of a significant delivery of power from the MENA countries in 2020, this can be guaranteed only by extending the capacity of peak-load power plants using natural gas, and later using gas from gasification of coal.

In Europe, the natural gas consumption will double as compared to the year 2000, but will then fall back to its original level after 2020, when an increasing proportion of imported power from solar-thermal sources becomes available, along with geothermic and hydroelectric power, the latter from Scandinavia. European sustainable-energy sources which could provide a secure capacity are unfortunately limited in terms of their potential. Therefore, the import of solar thermal power from MENA to Europe – which came under consideration through the efforts of the Desertec Foundation – will be indispensable, in order to combine with the capacity and fuel availability of gas-fired peak-load plants to provide a continually-secured, sustainable energy production capacity.

In MENA countries, concentrating solar thermal power plants represent the only sustainable source which is in fact in a position to supply the rapidly-growing demand for electric power. It can deliver both peak-load and base-load power as needed and is thus an important contributor to grid stability. Fluctuating power from wind and photovoltaic sources cannot be smoothed out here, as in Northern Europe, because in these dry countries there is no hydroelectric power from pumped-storage plants. According to our scenario, fossil-fuel plants will be used by 2050 only as a backup, to some extent also as supplementary power source for solar-thermal plants. This will reduce the combustion of these fuels to an acceptable level and will put a limit on the otherwise soaring costs of electric power. Fossil fuels will be employed to guarantee that power is always available, while their consumption will be strongly reduced by the use of sustainable energy sources.

To complement the mixture of sustainable power sources, an efficient backup infrastructure will be necessary: On the one hand, it must provide a secure generating capacity oriented to consumers' needs and based on gas-fired peak-load plants which can react quickly to power shortages; on the other, an efficient power grid must be established to allow the transmission of power from the most suitable production sites for sustainable sources to the major centers of power consumption. One possible solution is a combination of high-voltage DC transmission lines (HVDC) with a conventional AC grid.

Power Transmission over HVDC Lines

By 2050, according to our scenario, transmission lines with a capacity of 2.5 to 5.0 GW each must be able to transport around 700 TWh of solar energy per year from 20 to 40 different locations in the Middle East and North Africa to the major centers of consumption in Europe (Table 9), thus supplying ca. 15 % of the European power requirements. The cost of these imports is based on low production costs of ca. 0.05 €_{2000}-ct./kWh and a high flexibility for base-load, balancing power and peak-load operation (see below). Since our present AC grids would have overly high transmission losses at such high power levels and long transmission distances, it will be necessary to employ HVDC transmission lines. HVDC is available as a mature technology and is becoming increasingly important for the stabilization of large-area power grids. It contributes to the strengthening of equalization effects between local and distant energy sources and to containment of operational interruptions in larger power plants by utilizing back-up capacity from far away.

In mid-2010 in China, a nearly 1500 km long HVDC transmission line was put into operation. It connects several hydroelectric plants with the large cities Guangzhou, Hong Kong and Shenzen with a transmission capacity of 5000 MW. This corresponds to the output power of five large plants.

Power will be transported over sometimes long distances throughout Europe and the MENA countries and then fed into the conventional grids. Analogously to a motorway network, a future HVDC grid will have only a few inputs and outputs, which connect it to the conventional AC grids. In this analogy, the present AC grids are comparable to the local and municipal street networks. They will carry out the local distribution of electrical energy, just as they do now. The energy loss in the HVDC lines over a dis-

tance of 3000 km will be about 10 %, in contrast to more than 45 % for transmission over conventional AC lines.

There is a widespread misconception that for every wind farm or photovoltaic installation, a fossil-fueled back-up power plant of the same generating capacity must be built. In contrast, a model of the hourly time evolution of the energy supply systems of selected countries showed, in accordance with our scenario, that even without additional energy-storage capacity, the balancing power from the existing peak-load plants suffices to equalize fluctuating demands. This holds so long as the fluctuating portion of the sustainable sources is smaller than the existing peak-load capacity, which is the case in our scenario.

In fact, the need for conventional base-load power plants will decrease step by step as a consequence of the growing proportion of sustainable energy sources (Figure 9). Base-load power will be generated using coupled heat and power plants burning both fossil and biomass fuels, by

FIG. 8 | INSTALLED POWER GENERATING CAPACITY FOR EU-MENA

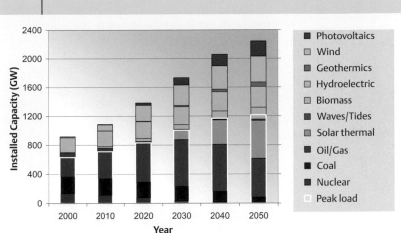

Installed generating capacity as compared to the cumulative peak load (white framing) for the entire EU-MENA region.

TAB. 1 | EU-MENA POWER EXPORTS FROM SOLAR-THERMAL PLANTS

Year		2020	2030	2040	2050
Capacity (GW)		2 × 5	8 × 5	14 × 5	20 × 5
Transfer (TWh/a)		60	230	470	700
Capacity utilization		0.60	0.67	0.75	0.80
Land area	CSP	15 × 15	30 × 30	40 × 40	50 × 50
km × km	HVDC	3100 × 0.1	3600 × 0.4	3600 × 0.7	3600 × 1.0
Cumulative	CSP	42	134	245	350
Investm. Bil. €	HVDC	5	16	31	45
Power cost	CSP	0.050	0.045	0.040	0.040
€ 2000/kWh	HVDC	0.014	0.010	0.010	0.010

Cumulative investments up to 2050 for power transmission lines and plants, and the overall power cost incl. transmission in constant Euros referred to the year 2000. Power cost was computed on the basis of 5 % interest and 40 years operating life for solar-thermal plants and 80 years for the HVDC power lines. CSP: concentrating solar power plant; Capacity: in each case the number of blocks in a power plant × output power in GW. Recently, an update of this scenario has been published in [9].

runoff-river hydroelectric plants and by wind and photovoltaic plants. Balancing power will be taken from more readily storeable sources such as hydroelectric plants with reservoirs, biomass or geothermics. This combination of energy sources will not completely meet the daily demand variations, but will approach them closely. The remaining peak-load capacity (or, more correctly, balancing power capacity) will be provided by pumped-storage plants, water reservoirs, solar-thermal plants and peak-load plants burning fossil fuels [6]. The generating capacity based on fossil fuels that is still in operation by 2050 will serve exclusive-

ly for load equalization or combined heat and power generation.

Cost-Effective Power from Sustainable Energy Sources

Through worldwide installation of solar-thermal power plants, the cost of solar power will be reduced at a progress ratio of 85 to 90 % each time the installed capacity doubles, due to learning and rationalization effects [7]. Here an example: A solar thermal power plant at present, depending on the insolation at its site, can generate power for 0.15 to 0.20 $€_{2000}$/kWh, assuming a capital interest of 6.5 % per year and a plant lifetime of 25 years. With 10,000 MW installed power capacity worldwide, this cost would decrease to around 0.80 to 0.10 $€_{2000}$/kWh, and with 100,000 MW installed capacity, to 0.04 to 0.06 $€_{2000}$/kWh. Similar learning curves can be observed for all of the sustainable technologies.

A cost reduction of this order would be achieved for an assumed expansion of solar-thermal generating capacity from today's 1000 MW to about 40,000 MW in the year 2020 and 240,000 MW by 2030, including the capacity needed for desalination of seawater. Current scenarios even assume notably faster growth rates for solar-thermal installed capacity. In the long term, an overall capacity of 500 to 1000 GW could be installed worldwide by 2050. All the costs quoted above are in constant (year 2000) Euro values, i.e. without inflation.

As soon as the point of cost equality with conventional power generating sources has been reached, sustainable sources would grow more quickly, thus avoiding further rises in national electric power costs. In this manner, the power cost of the energy mixture could be held constant, in some cases even reduced to an earlier level, by increasing the fraction of power from sustainable sources. This concept can be implemented in all the EU-MENA countries. The present continual escalation of power costs shows clearly that the broad-based introduction of sustainable energy sources offers the only possibility of avoiding further cost increases in the energy sector on a long term basis, and of returning to a relatively low level of electric power costs in the medium term.

An Alternative to Climate Change and Nuclear Power

Implementing our scenario could lead to a reduction of carbon dioxide emission values to a level which would be compatible with the goal of decreasing the carbon dioxide concentration in the atmosphere sufficiently to limit global warming to the range of 1.5° to 3.9° C. Assuming 1790 million tons of CO_2 emissions per year in the year 2000 for the EU-MENA region, these emissions could be reduced to 690 Mt/a by 2050, instead of increasing to 3700 Mt/a. The attainable level of 0.58 t/cap/a for the emissions per person and year due to electric power by 2050 is acceptable, in view of the recommended total emissions of 1.0–1.5 t/cap/a

FIG. 9 | GERMAN POWER GENERATION IN 2050

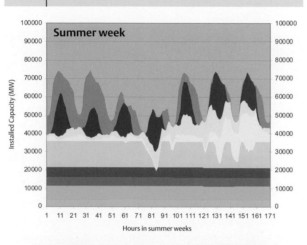

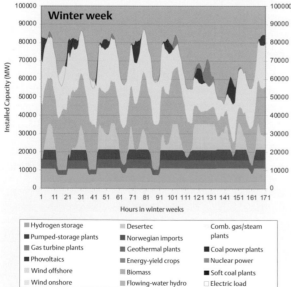

		Comb. gas/steam plants
Hydrogen storage	Desertec	
Pumped-storage plants	Norwegian imports	
Gas turbine plants	Geothermal plants	Coal power plants
Phovoltaics	Energy-yield crops	Nuclear power
Wind offshore	Biomass	Soft coal plants
Wind onshore	Flowing-water hydro	Electric load

Fluctuating and balancing power from various sources in a typical summer and winter week in Germany, from a model scenario with 90 % sustainable energy by the year 2050. Installed generating capacity: Wind power, 65 GW; Gas turbines, 60 GW; Photovoltaics, 45 GW; Solar imports, 16 GW; Biomass, 8 GW; Pumped-storage plants, 8 GW; Runoff-river hydroelectric, 6 GW; Hydroelectric imports from Norway, 6 GW; Geothermal, 5 GW.

given by the Scientific Council on Global Environmental Change (WBGU) of the German Federal government. Other harmful emissions will also be reduced by this scenario, without having to resort to the increased use of nuclear power, with its associated risks.

At present, the technology of capturing and storing carbon dioxide from coal-fired power plants is under development. This CCS process is treated in our study as complementary, but not as an alternative to sustainable energy sources. It in fact reduces the efficiency of the power plants and thus increases their consumption of fossil fuels by up to 30 %.

All in all, our scenario indicates a way of effectively reducing the negative environmental influences of energy production. The model is suitable for worldwide application, as confirmed by a study from the U.S. Department of Energy (DOE) on the feasibility of the concept in the USA [8].

In order to implement the project, the governments of the EU-MENA countries must take the initiative now and provide the legal and financial boundary conditions for investments in clean and sustainable energy. This is – not unimportantly – also a secure path to a sustainable water supply for the MENA countries.

Summary

Studies carried out by the DLR on the potential of sustainable energy sources in Europe, the Middle East and in North Africa have produced the following conclusions: Beginning with an existing fraction of 16 % sustainable energy sources in the year 2000, a fraction of 80 % could be achieved by 2050. To complement the sustainable energy sources, an efficient back-up infrastructure will be needed. It would yield a secure, demand-oriented electric power generating capacity with rapidly reacting peak-load plants burning natural gas and combining variable and flexible sustainable sources in a well-balanced way. Power transport to Europe would be carried out over high-voltage direct-current (HVDC) transmission lines. If a beginning power transfer of 60 TWh/a were installed between 2020 and 2030, it could be expanded to 700 TWh/a by the year 2050. The strong insolation in the MENA region and the low transmission losses of ca. 10 % by HVDC would lead to a competitive power cost of ca. 0.05 €$_{2000}$/kWh. Moreover, the quality of solar electricity imports on demand would be very high. Instead of the expected doubling of carbon dioxide emissions by the year 2050, this concept would reduce them to 38 % of their level in the year 2000. For the total sustainable power plant park in the EU-MENA countries, only 1 % of their land area would be required. This corresponds to the present-day land use in Europe for transportation.

References

[1] F. Trieb and U. Klan, *Modelling the Future Electricity Demand of Europe, Middle East and North Africa*, Internal Report, DLR **2006**.

[2] L. Mantzos and L.P.Capros, *European Energy and Transport Trends to 2030*, Update **2005**, European Commission, Brussels 2005: ec.europa.eu/dgs/energy_transport/figures/trends_2030.

[3] G. Benoit and A. Comeau, *A Sustainable Future for the Mediterranean*, Earthscan **2005**. Executive Summary: bit.ly/SSMDv4.

[4] S. Teske, A. Zervos, and O. Schäfer, *Energy (R)evolution*, Greenpeace, EREC **2007**: www.greenpeace.de/fileadmin/gpd/user_upload/themen/energie/energyrevolutionreport_engl.pdf.

[5] A.T. Kearney, *Solar Thermal Electricity 2025 – Clean electricity on demand: attractive STE cost stabilized energy production*, ESTELA, June **2010**.

[6] L.A.Brischke, *Model of a Future Electricity Supply in Germany with Large Contributions from Renewable Energy Sources using a Single Node Grid* (in German), VDI Fortschritt-Berichte, Reihe 6, Energietechnik, Nr. 530, VDI Düsseldorf **2005**.

[7] R. Pitz-Paal, J. Dersch, and B. Milow, *European Concentrated Solar Thermal Road Mapping*, ECOSTAR, SES6-CT-2003-502578, European Commission, 6th Framework Programme, German Aerospace Center, Cologne **2005**: www.promes.cnrs.fr/ACTIONS/Europeenes/ecostar.htm.

[8] H. Price, *DLR TRANS-CSP Study applied to North America*, U.S. Department of Energy (DOE) **2007**: www.nrel.gov/docs/fy07osti/41422.pdf.

[9] F. Trieb, C. Shillings, T. Pregger, and M. O'Sullivan, *Solar Electricity Imports from the Middle East and North Africa to Europe*, Energy Policy **2012**, *42*, 341–353.

About the Author

Franz Trieb has worked since 1994 in the Systems Analysis and Technology Evaluation Division of the Institute for Technical Thermodynamics of the German National Research Center for Aeronautics and Space (DLR). His main focus is on solar thermal power plants, solar energy resources, and the scenario for utilizing sustainable energy sources in Europe, the Near East, and North Africa.

Address:
Dr. Franz Trieb, Zentrum für Luft- und Raumfahrt, Institut für Technische Thermodynamik Pfaffenwaldring 38–40, 70569 Stuttgart, Germany. Franz.Trieb@dlr.de

Hydrogen: An Alternative to Fossil Fuels?

BY DETLEF STOLTEN

In times when nuclear power in some countries is being phased out and CO2 emissions must be reduced, the path towards renewable energy sources is clearly marked out. However, due to their strong fluctuations, renewable sources can not be very readily integrated into existing energy infrastructures. Can hydrogen serve as an energy-storage medium and provide a breakthrough here?

Hydrogen as an energy carrier was intensively discussed and investigated as early as the 1970's, due to the two oil price crises then. However, it soon became apparent that the technology was not mature enough for widespread applications and that it would not be cost competitive. The last hydrogen projects were discontinued in the early 1990's.

In the meantime, the situation has again changed. One aspect is the increasing proportion of fluctuating power generation from renewable energy sources such as wind and solar power. It is not sufficient, though, that merely the technological scenario for the use of hydrogen has changed; successful widespread application will also require a new socio-economic scenario. This chapter discusses both of these aspects.

Properties of Hydrogen

Hydrogen is the lightest element. Down to –253° C, it occurs in gaseous form as the H_2 molecule, while at still lower temperatures it is a liquid. The gas is completely non-toxic; this distinguishes it from e.g. gasoline, which is carcinogenic, damaging to genetic material, and a danger for ground water. Hydrogen has a broad flammability range of 4 – 75 volume percent in air; however it also has a very low density, so that it rises rapidly and is readily diluted. Hydrogen-fueled vehicles can thus be constructed to be generally safe and are certified for use in countries all over the world. They are already driving on the streets as licensed demonstration vehicles.

In nature on earth, hydrogen is not found in free form in technologically relevant quantities, since it reacts quickly to give water in our oxygen-rich atmosphere. Most of the hydrogen is present in oxidized form as water molecules. Furthermore, hydrogen plays an important role in the structures of organic molecules. It is found in chemically bound form in many organic compounds, not least in chemical energy carriers such as natural gas or gasoline and diesel fuels.

The Present-day Production of Hydrogen

Hydrogen is required for many chemical processes on an industrial scale and is produced to meet this demand; for example for the manufacture of fertilizers via ammonia, or to desulfurize and improve petroleum-based fuels by hydrogenation. This use of hydrogen as a chemical reagent is well established, including its manufacture, safe handling procedures, and cost factors. The main portion of this hydrogen is produced from natural gas via steam reforming. We give the reaction here using the example of the major constituent of natural gas, i.e. methane; heavier hydrocarbons react in an analogous manner. The chemical formula for the so-called steam reforming reaction is

$$CH_4 + H_2O \Leftrightarrow CO + 3\ H_2,$$

where CH_4 is methane, H_2O the well-known formula of water, CO is carbon monoxide, and H_2 is (molecular) hydrogen. The reforming reaction requires energy in the form of heat, namely 206 kJ (kilojoules) per mole, i.e. per 16 g of methane. The carbon monoxide contained in the reaction product gas is then removed by adding water vapor, initiating the so-called water shift reaction:

$$CO + H_2O \Leftrightarrow CO_2 + H_2,$$

in which 42 kJ of heat energy per mole is released (CO_2 is carbon dioxide).

Steam reforming yields a gas with about 71–75 % hydrogen, 11–18 % CO, 11–4 % CO_2, and a remaining methane content of 3–7 % [1]. This gas is further purified by the shift reaction down to ca. 1 % CO. Higher levels of purity can be obtained by pressure swing adsorption or membrane separation processes to give 99.999 % pure hydrogen. If especially pure hydrogen is required, it can be prepared by the electrolysis of water; it can then either be used directly or

Renewable Energy. Edited by R. Wengenmayr, Th. Bührke. Copyright © 2013 WILEY-VCH Verlag GmbH & Co. KGaA, Weinheim

further purified via membranes. In electrolysis, the energy required to decompose water molecules into hydrogen and oxygen is provided as electrical power. Hydrogen from electrolysis is pure; however, it is more expensive.

In the year 2002, worldwide $5 \cdot 10^7$ million tons of hydrogen as the pure substance and in hydrogen-containing gas mixtures were produced [2]. The major portion was used by refineries and for the synthesis of ammonia and methanol. Hydrogen production in 1991 was derived from natural gas (77 %) with a small proportion of oil; from coal (18 %); and from electrolysis (4 %) [3]. These fractions are still relevant today.

Aside from its use as a chemical reagent, hydrogen also has great potential as a chemical energy carrier and thus as a storage medium for fluctuating energy technologies. It must in any case be produced artificially. It is therefore not a primary energy source, but rather a secondary source. The main role of such sources is to serve as storage media for energy.

The New Scenario for Hydrogen

With regard to energy technology, hydrogen has become relevant again today due to environmental problems. The extensive combustion of fossil fuels such as coal, petroleum and natural gas is a large-scale intervention by mankind into the natural balance of the earth. The global warming provoked by the increasing concentration of CO_2 in our atmosphere will have threatening effects if a further rise in the average temperature cannot be kept within narrow limits. The EU has set an ambitious goal of 2° C above the preindustrial level for this rise [4]. Today, 0.8° of this EU goal has already occurred [4], so that many countries have established climate-protection programs. They support technical alternatives to an energy supply which releases large quantities of CO_2. The climate program of the German Federal government stipulates a broad-based energy supply, depending primarily on renewable energy sources and including very ambitious CO_2 reduction goals [4].

Hydrogen as an energy carrier plays a role today in two areas, owing to the increasing proportion of sustainable energies in use and the more stringent requirements for energy efficiency. It thus represents the bridge between stationary electric power production and the transport sector:

1. For transportation, automobiles and buses can be powered very efficiently and with no CO_2 emissions using hydrogen and fuel cells;
2. Many renewable energy sources, especially wind energy, but also solar energy, whose technical and economic feasibility are at present guaranteed, generate electric power in a fluctuating mode and therefore require energy storage systems. This storage could be accomplished by using part of the wind (and solar) power generated to produce hydrogen electrolytically, and then storing it.

These two elements can be coupled in a rational way by using so-called surplus power, e.g. wind power which is not needed by the power grid at the time when it is generated, for the production of hydrogen. This hydrogen can then substitute mineral fuels directly.

For heavy machinery such as trucks, construction equipment, locomotives, ships and aircraft, liquid fuels cannot be readily substituted, and this will continue to be true in the foreseeable future. For these uses, the energy storage density of hydrogen at the current and foreseeable state of the technology is insufficient in comparison to liquid fuels. For stationary applications also, gaseous hydrogen will not play a role in the foreseeable future; here, a mature infrastructure for utilizing natural gas is already in place.

Hydrogen could, however, be added at up to several percent to natural gas in the pipeline networks, as has been discussed under the buzzword "wind hydrogen". As an alternative, methane production from hydrogen has been suggested, where hydrogen and carbon dioxide are reacted to give methane with an energy input (thermodynamically more precisely: an uptake of enthalpy) of 206 kJ/mole:

$$CO_2 + 4\, H_2 \Leftrightarrow CH_4 + 2\, H_2O.$$

The carbon dioxide could be provided by a future carbon capture and sequestration (CCS) scheme. This alternative indeed makes use of an established infrastructure, but it is not energy efficient and requires the availability of concentrated and purified CO_2, e.g. through CO_2 capture from exhaust gases.

Methane production is in addition economically questionable, since pure electrolysis hydrogen used as a fuel would be in competition with it and would have at least the value of gasoline today, 65–70 €-ct./liter of gasoline equivalent before taxes. Even twice this price would be justified and therefore probably acceptable to the market, since hydrogen in fuel-cell powered vehicles is twice as energy-efficient as gasoline burned in an internal-combustion engine. Thus, one could expect that a price of 1.3–1.4 € per liter of gasoline equivalent would be achievable for hydrogen (gasoline has an energy content of 32 MJ/l; MJ = megajoules; l = liter). Hydrogen used for methane production and added to natural gas would be sold at the same price as the natural gas, i.e. around 1 €-ct./MJ or 32 €-ct./liter gasoline equivalent. Thus, the value of a pure electrolysis product would be reduced by half or even to only a quarter by a process technology which involves additional energy losses and added costs.

Hydrogen in the Gas Tank

As the lightest element, hydrogen has a high mass-specific energy storage density of 120 MJ/kg. This corresponds to about three times the storage density of diesel fuel, at 43 MJ/kg, referred to the so-called lower heating value. However, this physically favorable property of hydrogen cannot readily be applied in a technological setting. While an automobile tank for gasoline or diesel fuel weighs about 10 kg, a pressurized gas system to store hydrogen for a cruising range of 500 km at today's best technical level would weigh 125 kg (see also the diagram in the chapter on fuel cells, p. 140). The corresponding storage density of the tank plus fuel is only about 5 MJ/kg.

Even though only about 4 % of the original physical energy density of hydrogen remains, the comparison to storage batteries will later show that pressurized hydrogen is still about an order of magnitude better than the mass-specific energy storage density of batteries. This mass consideration is important, since vehicles must be continually accelerated and decelerated, so that additional mass leads to a higher fuel consumption. Electric vehicles, including fuel-cell powered vehicles, are however able to recycle 1/3 to 1/2 of their braking energy and store it in batteries on board (see the chapter on electric vehicles).

As a gas, however, hydrogen has a much lower volume-specific storage density than the liquid hydrocarbons. Liquid hydrogen has a storage density of 8.5 MJ/l; as a pressurized gas at 700 bar, it has 5.1 MJ/l. If one takes into account the necessary volume of auto fuel tanks of about 70 l for the much more elaborate storage of liquid hydrogen or of gas at 700 bar, both technologies would have roughly the same storage density of around 4 MJ/l. Compared to the volume energy density of diesel fuel of 36 MJ/l, this is only around 10 %; compared to ethanol or methanol, it is 20 or 25 %. The comparison to liquid hydrocarbons and alcohols shows that the limiting factor is the volume needed for hydrogen fuel storage. This volume consideration is also important, since the enclosed space in the vehicle should be available as far as possible for payload use.

In comparison to batteries, on the other hand, hydrogen exhibits very favorable storage densities. Assuming today's best technology, lithium-ion batteries yield a storage density of about 1.3 MJ/l or 0.5 MJ/kg. This value does not take into account that larger batteries also require a cooling cell

and a cooling system, which lowers their effective energy density still further. For a fair comparison, the fuel cells would also have to be included in the density computation for hydrogen fuel. They weigh ca. 1–1.5 kg/kW of installed power output and thus, for current vehicle designs, about 100–150 kg. Then a storage battery system for a present-day automobile is about five times heavier than a comparable hydrogen fuel cell system including the fuel tank.

For the realistic operation of a vehicle, the efficiency of energy use of the fuel must also be considered. In order to obtain average values, a driving cycle is defined, i.e. a mixture of urban and long-distance driving under differing traffic conditions. Fuels cells attain nearly twice the efficiency of internal combustion engines for the European driving cycle; currently, the value is 55 %, with expectations of reaching 60 %. This is due to the fundamentally high efficiency of fuel cells, especially at lower power output levels: In city traffic, vehicles in the main use only 10–20 % of their maximum power. This effect plays an essential role in the high efficiency of hydrogen fuel-cell powered vehicles in the driving cycle.

Even though hydrogen cannot attain the high storage densities of liquid fuels, these figures demonstrate that it still offers a realistic substitute for mineral fuels. Hydrogen-based, fuel-cell powered vehicles with electric drive exhibit the same flexibility in operation as conventional autos with internal-combustion engines. This includes their weight, the available inside space, and in particular also the ability to rapidly refill their tanks with hydrogen. Battery-powered vehicles, in contrast, require several hours to recharge, which makes it impossible to drive uninterruptedly over longer distances. This is one of the great advantages of the hydrogen fuel-cell technology as an environmentally-friendly mobile power source. Refilling the hydrogen tanks takes only about five minutes.

A safe and familiar procedure for refilling the tank is available. There are already a number of hydrogen filling stations, not only in Germany (Figure 1), but also around the world. Likewise, the basic choice of fuel has been decided: The filling stations are to offer pure hydrogen gas. In their vehicle designs, nearly all the manufacturers are planning on gaseous hydrogen at a pressure of 700 bar, or in some cases at 350 bar. Important manufacturers worldwide are *Diamler AG*, *General Motors* including *Opel*, *Honda*, the *Hyundai-Kia Group*, and the *Nissan-Renault Alliance*, as well as *Toyota* [5]. These vehicles are already at a mature stage of development and are expected to come on the market in 2015. Until then, essential development goals are price reductions for the fuel-cell systems and improvement of their operating lifetimes (see the chapter on fuel cells).

The Hydrogen Infrastructure: A Roadblock?

The question arises as to whether setting up a whole new infrastructure for using hydrogen as a motor fuel does not represent a serious obstacle to its widespread use. The answer is given by a study which was commissioned by the

Fig. 1 *This hydrogen filling station is a demonstration project for the Clean Energy Partnership* (www.cleanenergypartnership.de/en/news) (photo: CEP).

German Federal government to evaluate the marketability of fuel-cell powered vehicles, and thus paid particular attention to the infrastructure required [6]. A principal assumption of this study was that the infrastructure should be set up in several regions of high population density, where a sufficient number of vehicles could be concentrated. The hydrogen supply is to be provided using tank trucks for liquid hydrogen or by connecting the filling stations to the existing industrial pipeline network; pipelines transport gaseous hydrogen. As an accompanying development, the pipeline network in the chosen regions would be extended, and these local networks would be interconnected in the long term. The study estimated the overall cost, including the filling stations, to be around 120 billion €.

Figure 2 shows the layout of such a pipeline network based on this study. It corresponds to the so-called moderate scenario [6] and defines central feed-in points (filled circles on the map). The points on the North Sea and Baltic coasts are feed-in stations for electrolysis hydrogen from offshore windparks. The inland points are stations for feed-in of hydrogen produced from soft coal with CO_2 capture, using coal from the fields in the Rhineland and the Lausitz regions.

In this design, nearly all of the 10,000 filling stations in Germany are connected to the pipelines. The hydrogen pipelines would be built along existing natural gas pipeline routes. 12,000 km of main pipelines and 36,000 km of regional distribution lines would be required, at a cost of 22–25 billion €. For comparison, the high-speed train connection between Nuremberg and Munich will cost 3.6 billion €. A hydrogen pipeline network is thus not a major cost problem.

In order to lay out all together 48,000 km of pipelines within Germany, there will however have to be a great deal of effort expended to convince the public. This will require a persuasive and complete demonstration of the technology, making its advantages clear. This demonstration will be mainly based on liquid hydrogen, which is vaporized at the filling stations and compressed to the desired pressure before filling into the vehicle tanks. The liquid hydrogen will initially be delivered by tank trucks, just like conventional fuels. The liquefaction of the hydrogen to be sure exacts its price in energy and cost, corresponding to about 30 % of the energy content. The use of liquefied hydrogen is therefore inferior to pressurized gas, which can be distributed via pipelines; the latter involves an efficiency loss of only 10–15 %. But for the introduction of hydrogen onto the market, liquefaction is indispensable, just as it is for the long-term supply to distant sites where the construction of a pipeline would not be economically feasible.

And Where will the Hydrogen Come From?

Hydrogen can be produced from a variety of sources. The environmental-political requirement that it be produced from renewable sources limits the choice, and eliminates the current industrial production methods. Possible re-

FIG. 2 | A FUTURE PIPELINE NETWORK

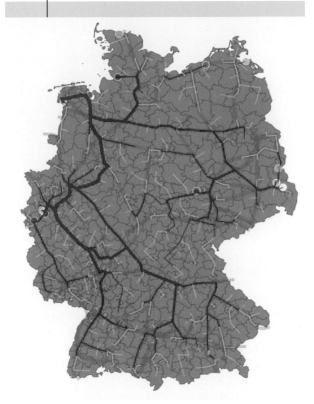

A possible future pipeline network in Germany, according To the 'moderate scenario' [6]. The circular points are feed-in stations.

newable energy sources for hydrogen production are wind power and photovoltaics; they would provide power for the production of hydrogen by electrolysis. Both of these energy sources fluctuate strongly: At certain times, they generate surplus power, while at other times, they cannot cover the power needs, even though their average overall power output would be sufficient for the total requirements. At present, photovoltaics deliver around 1 % (Vgl. Kohl: 3 % in 2011) of the electric power used in Germany, and wind power delivers about 9 % (Vgl. Kohl: 7,5 % in 2011). At the current state of development, wind power has a much greater potential of being able to supply an appreciable portion of the overall electric power consumption from a renewable source.

Power generation from wind energy coupled with hydrogen production via electrolysis has a strong potential. On the average, the power output of German wind-power plants is today about 1.2 MW, while the largest plants generate 7.5 MW of installed power [7]. If one considers the complete replacement of all the smaller plants by 7.5 MW plants (repowering) and the planned construction of offshore windparks with a total output of 35 GW, even without power generation using nuclear energy and anthracite and lignite coal, there would be an energy surplus of about 18 % or 91 terawatt-hours (TWh) per year. In this computation, we have left the current power generation capacity from natural gas unchanged. In order to compensate for the strong fluctuations in wind power, it was also assumed that households would require up to 50 % less natural gas in the future, since the houses will have improved insula-

FIG. 3 | FUTURE ELECTRIC-POWER GENERATION

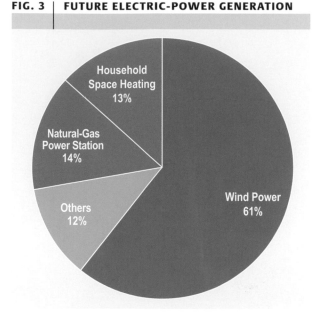

The distribution of future electric-power generating sources in Germany, based in large part on renewable energy sources. Not shown: Hydrogen produced from surplus wind power could make a major contribution to powering transportation.

tion. The proportion of energy used from natural-gas combustion would then be 14 % of current electric power consumption and 13 % of the current household space heating (Figure 3). We have assumed an efficiency of 58 % for electric power generation using natural gas.

Naturally, these large wind-power plants could not be set up with the same density and on the same sites as the current smaller plants. This estimate nonetheless shows that Germany could relatively easily supply all of its electric power needs using wind power. The problem with renewable energy sources lies, as mentioned, in their fluctuations. The gaps in wind power could be filled by natural-gas fueled plants: Both gas turbines and modern combined-cycle gas and steam plants have excellent dynamics and are suitable for meeting peak-load requirements.

The surplus energy from times of maximum wind power generation could, as we have discussed, then be stored in the form of hydrogen. Similarly to natural gas at present, it could be stored in technical or geological storage tanks in caverns or salt domes or in porous rock formations. Later, the stored hydrogen could be converted back into electric power using either gas power plants or fuel cells; this would however reduce overall CO_2 emission less than its use as a motor fuel. It is thus preferable to employ hydrogen in the transportation sector. Used as a motor fuel, hydrogen guarantees that no CO_2 will be emitted by the vehicle. Neither mineral fuels nor biofuels, including alcohols, can achieve this.

Let us assume, quite realistically, that a fuel-cell powered vehicle uses 1 kg of hydrogen per 100 km, corresponding to a diesel consumption of 3 l per 100 km. The hydrogen is presumed to have been produced by electrolysis with 70 % efficiency. Then, using the surplus wind power estimated above, 16 million vehicles with the current German average range of 12,000 km (Vgl. Vezzini, S. 2: fast 33.000 km/a Germany (90 km/d); ca. 11.000 km/a Schweiz

(30 km/d)) per year could be operated with this surplus power. This corresponds to 38 % of all German vehicles in the year 2010. Hydrogen could thus make a significant contribution to establishing an environmentally-friendly transportation system on the basis of sustainable energies.

Summary

The fight against climate change demands increasingly CO_2-free electric power generation. Most of the renewable energy sources which can provide this are however subject to strong fluctuations in their power production. It is thus reasonable to carry out electrolysis during periods of surplus power generation and to store the hydrogen produced. Vehicles powered by fuel cells can make use of this hydrogen with a high efficiency. This will permit operating a significant portion of the vehicles without CO_2 emissions and will make a sensible and economically favorable use of the surplus power from renewable sources. This example shows that in using renewable energy sources, the energy sectors will redefine themselves or will fuse. In the concept introduced here, natural gas that was up to now consumed for space heating will be used to provide controlling power for the grid, while wind power will be used indirectly via hydrogen in the transport sector.

References

[1] J. Rostrup-Nielsen and L.J. Christiansen, *Concepts of Syngas Manufacture*, Imperial College Press, London **2011**.

[2] J. Rostrup-Nielsen, Cattech **2002**, *6(4)*, 150.

[3] D. Woehrle, Nachr. Chem. Tech. Lab. **1991**, *39(11)*, 1256.

[4] European Commission, European Commission – Climate Action. Online: ec.europa.eu/clima/policies/brief/eu/index_en.htm.

[5] P. Froeschle, Fuel Cell Power Trains, in: *Hydrogen and Fuel Cells*, D. Stolten (ed.), Wiley, Weinheim **2010**, p. 793.

[6] GermanyHy, Woher kommt der Wasserstoff bis 2050, Bundesministerium für Verkehr, Bau und Stadtentwicklung, Berlin **2009**.

[7] Enercon website, Enercon aktuell, January 27th, 2011. Online: www.enercon.de/en-en).

About the Author

Detlef Stolten studied metallurgy and non-metallic materials at the Technical University in Clausthal-Zellerfeld and received his doctorate there while working at Robert Bosch GmbH. Following 12 years in industry, in 1998 he became Director of the Institute for Energy and Climate Research – Fuel Cells at the Jülich Research Center. Since 2005, he has been a member of the Strategic Council on Fuel Cells and Hydrogen of the German Federal ministries BMVBS, BMWI and BMBF. In 2010, he was chairman of the World Hydrogen Energy Conference, WHEC

Address:
Prof. Dr.-Ing. Detlef Stolten
Forschungszentrum Jülich GmbH,
Institut für Energie- und Klimaforschung (IEK),
IEK-3:Brennstoffzellen
D-52425 Jülich, Germany.
d.stolten@fz-juelich.de

Seasonal Storage of Thermal Energy

Heat on Call

BY SILKE KÖHLER | FRANK KABUS | ERNST HUENGES

In energy-supply systems with combined heat and power (CHP) generation, or also in connection with sustainable energy sources, especially solar energy, the question often arises as to how to store the thermal energy. This can involve short storage times such as hours or days, but it continues on up to annual (seasonal) heat storage.

Solar radiation and wind are – as everyone knows – not available around the clock; instead, they occur at variable times and with a limited predictability. Only through the use of energy storage systems can these renewable energy sources be made reliable and readily available for energy provision. Electrical energy can, for example, be converted into chemical energy – the buzzword here is the hydrogen economy (see the previous chapter) – and thus stored. In the case of thermal energy, in contrast, the seasonal variations make it attractive to save up the excess heat from the summer for use in winter, or conversely, to use the cold of winter for cooling in the following summer.

The same is true of installations for combined heat and power production (CHP). They make use of most of the thermal energy for both heating and electric power generation. In this way, rather high efficiencies can be achieved using CHP. The deployment of such installations has to be oriented towards the end-user energy requirements, i.e. electric power or heat. Accordingly, one refers to heat-directed operation (the determining parameter is the requirement for heat) or power-directed operation (determining parameter is the requirement for electric power) of the CHP installation.

Most installations are operated in the heat-directed mode, since the electric power not needed by the local user can always be fed into the power grid, which acts as a large storage system. Since, however, the price of electrical energy is much higher than that of heat energy, it would often be preferable from the economic standpoint to operate the installation in power-directed mode. The operator could then compensate expensive peak loads by producing more electric power, for example. If the heat produced at the same time is not needed by local users, it has to be re-leased as waste heat to the environment, or, preferably, stored. This excess heat can then be stored in a seasonal storage system. Storage of excess thermal energy is thus in the end a precondition for the energetically and economically expedient power-directed operation of CHP installations.

The Thermodynamics of Energy Storage

Such a storage system consists essentially of four functional units: the storage medium, the charging and discharging systems including the heat-transport medium, the storage container structure, and auxiliary systems.

In order to classify systematically the types of storage, the well-known distinction between open and closed systems, defined in thermodynamics, is useful here. In open systems, energy and matter can be transported across the system boundary. In closed, non-adiabatic systems, only energy and not matter can be transported.

The corresponding artificial or geological structures are popularly referred to as heat or cold reservoirs. This is problematic, since heat is only a transport quantity, which can pass across a system boundary, and in this sense it cannot be stored. Like work, it is a process quantity and is therefore dependent on the path of a process. The heat transported across the boundary of the storage system changes the internal energy of the storage medium, which in most cases is reflected in a change of the temperature of the medium. It would thus be more physically correct to speak of the storage of thermal energy.

Whether a storage system serves as a source of heating or of cooling in an energy-supply installation is in the first instance unimportant for the construction of the system. Thus, some thermal storage systems are used in alternate modes, and supply cooling in summer and heating in winter. To be consistent, we will refer in the following to charging of the storage system when heat is fed into it, and to discharging when heat is extracted from the storage system. Since these terms have become common, we will continue to refer to heat reservoirs (when the system is mainly used as a source of heat) and cooling reservoirs (when it mainly serves as a source of cooling or as a heat sink).

Considering the energy balance with the surroundings of a heat reservoir, its interactions with these surroundings and its losses can be specified. The system boundary contains the storage medium, the charging and discharging system, and the heat transport medium. The sum of the heat

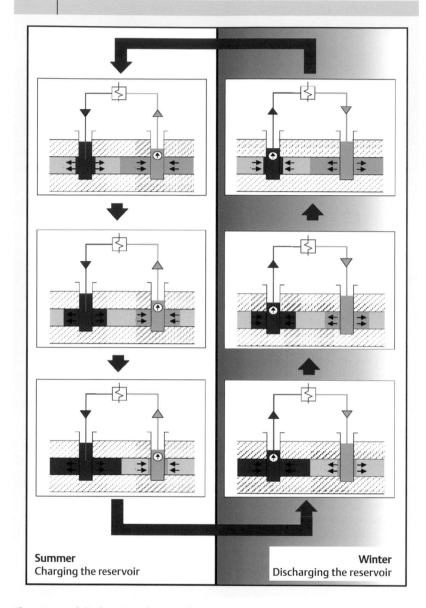

FIG. 1 | AN AQUIFER STORAGE SYSTEM

Summer
Charging the reservoir

Winter
Discharging the reservoir

Charging and discharging of an aquifer storage system (source: GTN).

also be considered. These occur as a result of equilibration processes within the storage system, such as a flow between the layers in a stratified hot-water reservoir. These losses can be included in the exergy balance around the reservoir. The exergy is that portion of the energy which can be unrestrictedly converted into any other form of energy. The change in the exergy in a storage system is the sum of the exergies input into the system and output by the system across its boundary (charging and discharging plus exergy losses accompanying heat losses), in addition to a conversion term. The latter takes into account the annihilation of exergy, which for example occurs on mixing within stratified hot-water storage systems by convection, or through thermal conductivity.

A goal in the construction of storage systems is of course to reduce the losses as far as possible, i.e. to improve the storage efficiency. It will be seen, however, that in designing different types of seasonal storage systems, such as aquifer and hot-water reservoirs, quite different concepts are required.

Aquifers as Seasonal Storage Systems

The technology of aquifers as energy storage systems is in many respects closely related to hydrothermal-geothermal energy. The storage medium is the natural substratum, i.e. the rock layers and the deep water they contain; the latter also serves as heat-transport medium. Aquifer storage systems are as a rule accessed via two boreholes or groups of boreholes. These are placed at a certain distance from each other, in order to avoid mutual thermal influences. The systems are open below ground and closed above ground. Above ground is a heat exchange apparatus, so that only energy transport – and no matter transport – occurs.

Both boreholes are fitted out with pumps and injection piping, which allows the flow of the heat-transport medium in the above-ground part of the plant to pass through in either direction. For charging, that is for storing thermal energy in the reservoir, water is taken from the cooler boreholes, warmed by the above-ground apparatus, and injected into the warm borehole. For discharging, the direction of flow is reversed: The pump in the warm borehole extracts water up to ground level, where it can give up heat to the heat exchanger system. The heat transferred is proportional to the mass flow rate of the thermal water in each case.

Charging and discharging of the aquifer storage system takes place horizontally (Figure 1). Around the warm borehole, a heated volume of water forms, which is pumped out again on discharging. Since the thermal water serves both as heat-transport medium and as storage medium, the temperature present in the reservoir is also available above ground, as long as the thermal losses within the borehole can be neglected. The discharging temperature will, however, always be lower than the charging temperature, since the warm water volume cools at its outer surface. This heat transport to the cooler surroundings takes place through

input and output and the heat losses equals the change in the internal energy of the storage system. The quantity of heat input or output is equal to the temperature change of the storage medium multiplied by its heat capacity, i.e. the product of its volume, density and specific heat.

Owing to the temperature change of the storage medium, there is a temperature difference relative to its surroundings. This temperature difference leads to storage losses. A closed storage system for example loses thermal energy via heat conductivity across the container walls. In an open storage system, heat can also be transported across the boundary of the system together with a mass flow.

Along with such external storage losses through interactions with the surroundings, internal storage losses must

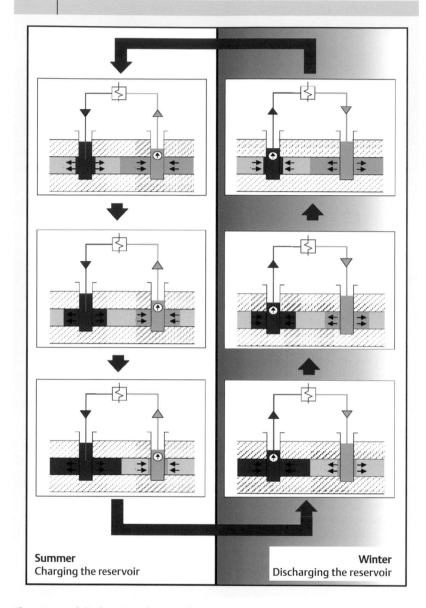

heat conduction and natural convection, as the warm water mixes with the cooler water from the surroundings, and through groundwater flows.

From this, we can derive the initial requirements for aquifers which are to be used for seasonal energy storage:

- The geological formation should be closed above and below, and the natural flow rate of the ground water as low as possible (preferably zero), to prevent the warmed or cooled water from flowing away.
- The temperature below ground increases as a rule with increasing depth. Heat storage reservoirs will thus be more readily found in deep strata than cooling reservoirs, since then the natural temperature of the aquifer lies nearer to the desired mean storage temperature, and storage losses are reduced. The requirement that the natural aquifer temperature be in the range of the mean storage temperature cannot always be fulfilled in an economically feasible way, in particular for high-temperature storage. A rough calculation shows: In order to obtain a temperature of 70 °C below ground, given an average temperature increase with depth of 30 K per kilometer, and 10 °C external temperature, boreholes of two kilometers depth would be required.
- To keep the required pumping power low, which is desirable, a high water throughput in the horizontal direction within the rock layers is desirable.

The maximum storage temperature and the storage volume, which determine the total storage capacity, are given by the natural properties of the aquifer. An advantage of aquifer storage is its relatively low investment cost. It is about 25 €/m³ (including planning, excluding taxes, for storage volumes of more than 100,000 m³) [1].

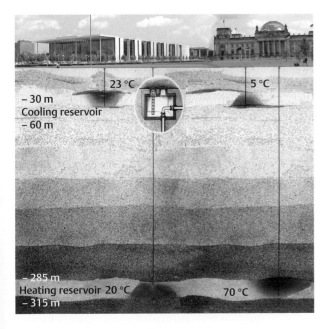

Fig. 2 An aquifer reservoir for heating and cooling *(source: GTN, BBG).*

FIG. 3 | INPUT AND OUTPUT RESERVOIR TEMPERATURES

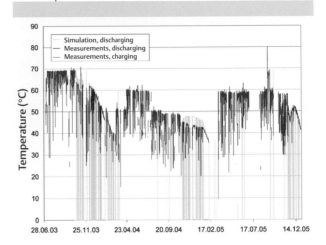

Charging and discharging temperatures of the thermal storage medium. Red and dark blue lines: measurements; light blue: model calculations.

Aquifer Reservoirs for Energy Management – Reichstag and Neubrandenburg

In the construction and the energy management of the new government buildings along the River Spree in Berlin, future-oriented, environmentally responsible and exemplary energy concepts were required [2]. The energy-supply system of the Berlin parliament buildings therefore contains, along with components for combined heat-cooling-power production, two aquifer storage reservoirs (Figure 2). The heat storage reservoir increases the contribution of the combined heat and power plant to the overall energy supply, even when the system is power-directed. The cooling reservoir permits the use of the low winter temperatures for air conditioning in the summer months.

The storage region is a sandstone stratum at a depth of 285 to 315 m which contains salty water (brine). Two boreholes access this aquifer. It is charged as a rule with water at a maximum of 70° C, and discharged at 65–30° C. In operation, a volume flow rate of up to 100 m³/h can be extracted or injected. The maximum charging power is about 4.4 MW. Heat from the discharge supplies the low-temperature sectors of the various building heating systems via direct heat exchange. Additional cooling of the water (down to a minimum of 20° C) can be carried out using absorption heat pumps as needed; their installed cooling power is ca. 2 MW.

This project is unique and has been accompanied by extensive scientific monitoring. Among other things, numerical models of the substrata have been constructed within the framework of research projects; they allow the temperature evolution in the storage media to be predicted [3, 4]. Figure 3 shows as an example the temperature at the wellhead of the warm borehole during the time period from June 2003 to December 2005. This period includes almost three charging and discharging cycles. The decreasing output temperature during discharging is characteristic of aquifer storage. It results from the losses described above, due to thermal conduction and convection at the perimeter of the warm storage volume.

At a considerably shallower depth of 50 m, under the bend of the Spree, another aquifer storage system was developed. It is used mainly for cooling of the buildings. In operation, fresh ground water is cooled in winter to 5° C. This is carried out essentially at outside temperatures below 0° C in dry cooling towers by heat exchange with the surroundings [1]. In the summer, this cooling reservoir supports the air conditioning systems via direct heat exchange.

In Neubrandenburg, a city in the German state of Mecklenburg-Vorpommern, a major portion of the buildings are connected to a central 200-MW district heating network. Its base load is carried by a combined-cycle gas and steam turbine power plant with 77 MW of electric output power and 90 MW thermal power. Since heat supplies in Neubrandenburg in the summer are essentially used only for providing warm water, the thermal power consumption is relatively low. As a rule, it is considerably less than the heat which is produced even when electric power is being generated at minimum load. The difference of up to 20 MW was in the past simply released to the environment through a cooling tower. Today, part of this 'waste heat' is fed into an aquifer storage system and used for heating in the winter in a portion of the district heating network which operates at a relatively low temperature level. The thermal power in this network is 12 MW at 80° C input temperature and 45° C return temperature.

The aquifer storage system consists of a cool and a warm borehole spaced ca. 1300 m apart. Both boreholes tap the operating stratum at a depth of about 1200 m, and they can extract or inject 100 m³/h of thermal water. The natural temperature of the stratum at this depth is about 55° C [10].

In summer, cooler water is pumped out at 40 to 50° C and is stored underground at a temperature of 80° C. In win-

FIG. 5 | GEOTHERMAL PROBES

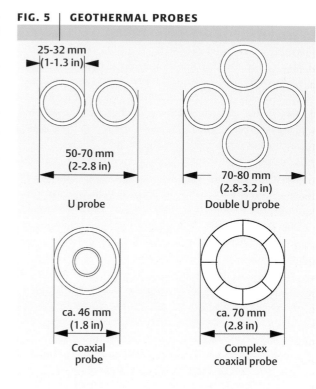

Cross sections of typical geothermal downhole heat exchangers [9].

ter, the flow direction is reversed; water is now extracted from the warm borehole. The output temperatures are between 80° C and 65° C, depending on the time until the water is extracted.

According to plan, it is expected that in the months April to September, a quantity of thermal energy equivalent to 12,000 MWh can be stored in the reservoir. In the winter, 8,800 MWh of this can be recovered through direct heat exchange at a thermal power of 4.0 to 2.9 MW.

The Neubrandenburg aquifer storage system has now been operating for four regular annual cycles. Figure 4 shows as an example the operating range of the reservoir during charging.

Storage in a Field of Downhole Heat Exchangers

For the storage of thermal energy in a field with downhole heat exchangers (DHE), the underground strata also serve as the storage medium. Here, the reservoir is tapped by a number of boreholes of 20 to 100 m depth, usually symmetrically arrayed, and these are outfitted with probes.

A geothermal field with DHE is an underground closed system in which there is no hydrological connection between the storage medium (earth) and the heat transport medium. The heat transport for charging and discharging takes place through the walls of the probes. The charging and discharging thermal power is thus proportional to the

FIG. 4 | EXCESS HEAT

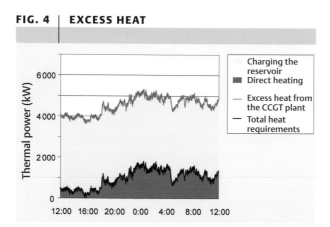

Excess heat and its use on the 5th and 6th of November 2005 in the Neubrandenburg aquifer reservoir.

surface area of the probes and to the temperature difference between the thermal transport medium and the storage medium. The cross-sectional area of the probes is therefore dimensioned in such a way that the probe surface area is as great as possible.

Figure 5 shows four types of probes in cross section. The U- and double U-tube probes consist of two or four tubes, respectively, which are connected together at the bottom of the borehole. In coaxial probes, the medium flows downwards in the outer space and upwards in the inner tube. A water-glycol (antifreeze) mixture is used as heat-transport medium, so that the liquid does not solidify at low temperatures.

Owing to the temperature difference between the heat-transport medium and the storage material, which is required for heat transfer, the temperature in the storage reservoir is always higher (when heat is being extracted from the reservoir) or lower (during charging) than the temperature of the heat-transport medium. These losses appear in the storage balance as exergy annihilation. The heat-transport medium in a DHE field storage system has a smaller temperature range in comparison to that in an aquifer storage system. Therefore, in energy-management systems with DHE field storage, heat pumps which raise the heat to a usable temperature level are usually employed.

The heat-transport medium flows through the heat-exchanger tubes and, during winter operation, takes on heat from the surrounding earth. In summer operation, it gives up heat to the earth and thus recharges the storage reservoir. In contrast to aquifer storage systems, the heat-transport medium always flows through the DHE in the same direction. The workable temperature variations and the storage capacity of the borehole field are dependent on the composition of the subsurface earth and on the heat-transfer properties of the tubes and their thermal contact with the earth. The reservoir cannot be isolated from its surroundings below ground. In order that the stored thermal energy not be lost, such reservoirs are usually located at sites where the ground water has only a very low – or zero – flow velocity.

Example: The Max-Planck Campus in Golm

In Golm, near Potsdam, in 1999 the Max Planck Society founded a new science campus, comprising three institutes. The energy management system is based on a combined heating-cooling-power system, making use of a geothermal borehole field for storage (Figure 6). This concept includes a cogeneration plant driven by an internal-combustion engine; the excess heat is used for space heating and warm water.

The DHE field contributes to cooling in the summer months, and in the process it stores heat energy. In the winter, it serves as thermal reservoir for a heat pump to supply heat energy to the buildings, and is thereby cooled, causing the temperature in the borehole field to sink below that of the surrounding earth. The field consists of 160 boreholes, each with a depth of 105 m, and it occupies an area of 65 m × 50 m, with an earth volume of about 400,000 m³. Its overall storage capacity is 2.24 MWh, and the input/output power is nominally 538 kW.

The temperature is measured year-round at four of the boreholes. Three of these are within the field, while the fourth gives values from the undisturbed region outside the field for comparison. Each borehole contains four temperature probes at depths of 15 m, 40 m, 70 m, and 100 m, respectively. The measured temperatures during the time period from September 2001 to September 2002 in one of the boreholes within the field are shown in Figure 7. Normally, the natural temperature increases with increasing depth. During charging, the warm heat-transport medium flows downwards and gives up its heat to the storage medium; thus, the temperature gradient in the upper part of the bore-

Fig. 6 *DHE field beneath the Max-Planck Campus in Golm (source: H. Jung).*

FIG. 7 | THE BOREHOLE FIELD IN GOLM

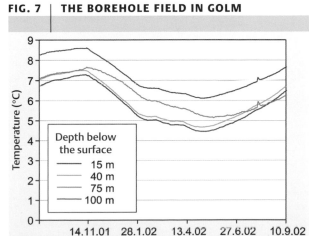

Measured temperatures from September 2001 to September 2002 within the bore-hole field of the Max-Planck Campus at a depth of 15 m, 40 m, 70 m and 100 m.

FIG. 8 | A HOT-WATER STORAGE SYSTEM

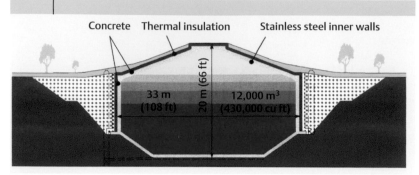

Schematic of the hot-water reservoir in Friedrichshafen, Germany (graphics: ilek, Univ. Stuttgart).

hole may be reversed after charging. In the lower part, it maintains its original sense at all times.

Hot-Water Storage

Hot-water storage systems are the most widespread type of reservoirs for thermal energy. They are used for example in solar-thermal plants for short-term intermediate storage. The reservoir is filled with hot water which is heated (charging) or cooled (discharging) by a heat exchanger system. The storage medium remains within the system in normal operation; only heat is transferred across the system boundary. These reservoirs thus represent closed, non-adiabatic systems with respect to their surroundings.

However, hot-water storage systems are operated not only with pure water. Aqueous solutions such as brines can be used, and even additional storage media like gravel can be employed. This generalization allows a better systematic use of the typical structural forms, tanks and basins. A common characteristic is that they are always artificially constructed.

Hot-water storage systems in the form of tanks are to be found in all sizes in energy supply installations, from short-term daily storage up to seasonal storage systems. For the latter, usually large cylindrical containers are employed, which can either be buried underground or set up above ground. Storage reservoirs at ground level or above must be able to withstand the environmental conditions of their surroundings and be adapted to them. Their walls are usually fabricated of reinforced concrete and are thermally insulated from their surroundings. Thermal losses occur only through heat conduction to the surroundings.

Water serves as the heat-transport and the storage medium. Due to this identity, the storage temperature of the energy system is available at its output as the usable operating temperature, as in aquifer storage systems. Within the reservoir, the temperature decreases from above to below. On discharging, the water with the highest temperature at the uppermost point of the reservoir is extracted, cooled in the energy-transfer system, and then returned to the low-

est, coolest region in the reservoir. This temperature stratification, which normally will already appear as a result of density differences, will be affected by thermal conduction and convection. While thermal conduction within the reservoir cannot be avoided, convection can be held to a minimum. Thus, for example, mechanisms for storing the thermal energy within the layer of the same temperature can be useful. This is especially interesting for solar heat, which can be stored at varying temperatures.

Since the storage reservoirs are usually operated at ambient pressure, the maximum storage temperature lies below 100 °C. In contrast to aquifer storage systems, they have a clear and well-defined storage capacity. The investment costs of these storage systems are estimated to be in the range of 450–120 €/m³ (depending on the overall volume) for concrete reservoirs, and 3000–600 €/m³ (for 0.2–100 m³ volume), or below 100 €/m³ (at volumes larger than 10,000 m³) for steel reservoirs [1].

Solar-Assisted Local Networks

The German Federal ministries for Commerce and Technology and for the Environment, Nature Protection and Nuclear Safety have subsidized the construction of pilot and demonstration plants for local solar heating networks within the framework of the program 'SolarThermal2000' [5–7]. In three of the subsidized pilot plants, in Friedrichshafen, Hamburg, and Hannover, seasonal hot-water reservoirs are integrated into the energy supply system as tank structures. In these solar-assisted local heating networks, use is made of the stored energy to preheat the return flow in the local heating network. If the heat output power or the temperature from the storage reservoir is not sufficient, heat from fossil fuels (gas, oil) or from district heating networks is used to reach the desired temperature.

The largest of these hot-water storage reservoirs is in Friedrichshafen, Germany. It has a height of 20 m and an inner diameter of 32 m, and thus provides a storage volume of 12,000 m³. In the final construction stage, this local heating network will supply heat to 570 dwellings with a heated floor area of nearly 40,000 m² (Figure 8).

The experience with the pilot projects for solar-assisted local heating with long-term energy storage has thus far yielded a fraction of solar heat in the overall energy consumption of 30 to 35 %. With additional improvements to the systems technology, solar fractions of 50 to 60 % are expected [8].

Gravel-Water Storage Systems

Gravel-water storage reservoirs are likewise artificial structures. In them, a mixture of gravel and water serves as storage medium, with a gravel concentration of 60–70 vol. %. This mixture is placed into a cavity in the ground which has been lined with a watertight plastic sheet. The maximum storage temperature is limited by the temperature stability of the plastic, and is typically in the range of 80° C. Due to the lower specific heat of the gravel, a gravel–water storage reservoir requires about 50 % more volume than a pure water storage reservoir for the same storage capacity.

Charging and discharging can be accomplished either by direct water flow through the reservoir or by heat exchange using coiled tubes. The heat-transport medium is thus either the water within the reservoir, or a second medium such as brine or an antifreeze mixture. In gravel-water storage systems, vertical temperature stratification is likewise observed, and it can be enhanced by the charging and discharging processes.

The gravel within the reservoir has two advantages: It supports part of the static load on the container, and thus makes its construction lighter and simpler. In addition, it reduces the free convection of the fluid within the reservoir and thereby the internal losses. The external losses are limited by thermal insulation applied outside the watertight sheet. In the local heating networks in Steinfurth and Chemnitz, which were also subsidized by the program mentioned above, gravel –water storage systems were utilized.

Summary

Energy from renewable sources is often not continuously available. The storage of thermal energy is therefore of great importance. Which type of storage technology is chosen for a particular application depends on various conditions. Seasonal storage can be applied successfully using structures such as hot-water storage in tanks and in gravel-water reservoirs. For some time, underground storage has also been practiced, using borehole fields or aquifers as storage media. The natural substratum is a complex storage medium, which requires a considerable effort for successful operation. However, such storage media can be developed with a considerably lower specific investment cost than hot-water storage systems.

References and Links

[1] BINE, Aquiferspeicher für das Reichtagsgebäude. 13/03, Fachinformationszentrum Karlsruhe, Bonn **2003**.

[2] B. Lützke, Energieversorgung für das Parlaments- und Regierungsviertel im Spreebogen Berlin, Bundesbaugesellschaft Berlin mbH, Berlin **2001**.

[3] S. Kranz, Energieversorgung der Berliner Parlamentsbauten – Integration der Aquiferspeicher in die Wärme-und Kälteversorgung der Berliner Parlamentsbauten. VDI Fachtagung "Geothermal Technologies", Potsdam **2008**, p. 199.

[4] S. Kranz and J. Bartels, Simulation and data-based optimization of an operating seasonal aquifer thermal energy storage. Proceedings, World Geothermal Congress, Bali (Indonesia) **2010**.

[5] BMU, Solarthermie2000, www.solarthermie2000.de.

[6] BMU, Solarthermie2000plus, www.solarthermie2000plus.de.

[7] M. Bodmann et al., Solar unterstützte Nahwärme und Langzeit-Wärmespeicher. Förderkennzeichen 0329607F, Solar- und Wärmetechnik Stuttgart (SWT), Stuttgart **2005**.

[8] BINE, Langzeit Wärmespeicher und solare Nahwärme, profiinfo I/01, Fachinformationszentrum Karlsruhe **2001**.

[9] M. Kaltschmitt, E. Huenges, and H. Wolff, *Energie aus Erdwärme*, Deutscher Verlag für Grundstoffindustrie; **1999**.

[10] BINE, Aquifer speichert Überschusswärme aus Heizkraftwerk, Projektinfo 04/07 Fachinformationszentrum Karlsruhe **2007**.

Note: BINE Informationsdienst is a service of the Karlsruhe Technical Information Center (FIZ) Karlsruhe GmbH: www.bine.info.

About the Authors

Silke Köhler, engineer for energy and process technology, heads the Research and Development at RWE Innogy GmbH. She has been working in the field of sustainable energy since 1996, among other places at the Institute for Solar Energy Research in Hameln/Emmerthal and at the GeoForschungsZentrum in Potsdam.

Frank Kabus studied thermal and hydraulic mechanical engineering at the TU Dresden; he obtained his doctorate at the TU Dresden on working materials for cyclic heat pumps. Since 1987, he has worked in the area of geothermal energy supplies. He is the general manager of Geothermie Neubrandenburg GmbH (GTN).

Ernst Huenges, physicist and process engineer, is head of the International Geothermics Center and the section Reservoir Technologies at the Helmholtz Center Potsdam GFZ – German Research Center for Geosciences. Currently, he is chairman of the research program Geothermal Technology of the Helmholtz Association and the leader of various geothermics projects.

Addresses:

Dr.-Ing. Silke Köhler, RWE InnogyGmbH, Gildehofstr. 1, D-45127 Essen, Germany. silke.koehler@rwe.com

Dr.-Ing. Frank Kabus, Geothermie Neubrandenburg GmbH, Seestraße 7A, D-17033 Neubrandenburg, Germany. gtn@gtn-online.de

Prof. Dr. Ernst Huenges, Helmholtz-Zentrum Potsdam, Deutsches GeoForschungsZentrum GFZ, International Centre for Geothermal Research, Telegrafenberg, D-14473 Potsdam, Germany. huenges@gfz-potsdam.de.

Fuel Cells for Mobile and Stationary Applications
Taming the Flame

BY JOACHIM HOFFMANN

Fuel cells permit clean and resource-efficient energy conversion. Nevertheless, this technology has developed only very slowly since its introduction in the year 1838 by the German chemist Christian Friedrich Schönbein. The much-publicized interest shown by the automobile industry has awakened fuel cells from their long sleep.

The German chemist Christian Friedrich Schönbein discovered the basics of fuel cells as early as 1838. Since then, this technology has developed only very slowly, although Jules Verne recognized over a hundred years ago that, "Water is the coal of the future. The energy of tomorrow is water which has been decomposed by electrical current. The elements resulting from this decomposition, hydrogen and oxygen, will secure the energy supply of the earth in the foreseeable future" [1].

Nevertheless, it was only the much publicized interest of the automobile industry which awakened the fuel cell

from its long sleep. Several well-known automobile manufacturers have demonstrated the technical feasibility of fuel-cell powered vehicles in the past decade. The driving forces behind this return to the fuel cell are in particular the growing scarcity of fossil energy resources and the increasing concentration of carbon dioxide in our atmosphere. Fuel cells consume less fuel per kilowatt hour or kilometer driven, owing to their high electrical efficiencies, and they emit correspondingly less carbon dioxide. Other environmentally harmful substances such as nitrogen oxides and soot particles can also be reduced or entirely avoided by the use of fuel cells.

If fuel cells are operated using pure hydrogen as fuel, the only "waste product" they release into the environment is water. Since hydrogen is however not available everywhere, and its storage is not simple (see the chapter "Hydrogen – an Alternative to Fossil Fuels?"), attempts are underway to adapt the fuel cells to operate with more readily handled fuels such as methanol, ethanol, gasoline or natural gas.

The operating principles and properties of various types of fuel cells are described separately in the infoboxes "How do Fuel Cells Work?" on p. 143 and " Types of Fuel Cells" on p. 145 [2–4].

Applications of Fuel Cells

Fuel cells have proven their reliability for decades in space applications, which provided the stimulus for their continuing technical development. They became sadly notorious during the flight of Apollo 13, when a leak in an oxygen tank not only endangered the supply of breathing air for the crew, but also shut down the electrical system [4]. Another field of application, which is now well established, is as power sources for submarines. Both applications are based on very stringent requirements on the reliability of the fuel cell systems. They are thus niches which have a negligible effect on the overall energy economy, in comparison to the widespread applications of fuel cells for stationary, mobile, and portable power which are currently being planned.

For stationary operation in power plants, fuel cells with power outputs in the range of 1 kW up to 10 MW are conceivable. They thus could provide decentral electric power for users ranging from single-family dwellings up to industrial plants. For these applications, most developers envisage the use of the so-called high-temperature systems: SOFC (Solid Oxide Fuel Cell) or MCFC (Molten Carbonate Fuel Cell). For installations in the range of 10 MW, in which

THE STORAGE CAPACITY OF VARIOUS ENERGY STORAGE DEVICES

System Fuel	System Fuel	System Cell
43 kg 33 kg	125 kg 6 kg	830 kg 540 kg
46 l 37 l	260 l 170 l	670 l 360 l

Comparison of the masses and volumes of various fuels, in each case for a cruising range of 500 km. The transparent figures represent the weight (above) and the volume (below) of the complete energy-storage system. Gaseous hydrogen (center) is under 700 bar pressure; the data are for electric-drive vehicles using fuel cells. The 6 kg of H_2 required contain 200 kWh of energy and thus twice as much as the lithium-ion batteries (right). The efficiency of the batteries for charging and discharging is considerably higher than the energy conversion in the fuel cells (see also the following chapter on electricvehicles) – but at the price of a much greater weight (graphics: Adam Opel AG) [7].

Renewable Energy. Edited by R. Wengenmayr, Th. Bührke. Copyright © 2013 WILEY-VCH Verlag GmbH & Co. KGaA, Weinheim

sufficient surplus heat at high temperatures is available for use in a subsequent gas and steam turbine combination (combined cycle, CC), electrical system efficiencies of over 85 % are conceivable. Processes that make use of the "waste" heat are for example steam generation or other cycles such as the Organic Rankine Cycle (see the chapter "Energy from the Depths"), a so-called decompression turbine, or also heating-cooling conversion machines (heat pumps).

An increase in the working pressure leads to a higher operating voltage for fuel cells. This effect is utilized in pressurized SOFC installations. In addition, the pressurized exhaust gases can drive a decompression turbine on expanding, allowing additional electrical power to be generated. Such hybrid systems of SOFC cells and turbines lead to overall thermal efficiencies of ca. 70 %. In comparison: The most modern power-plant technology today, a CC gas turbine and steam plant with several 100 MW output power burning natural gas, attains an efficiency of "only" 60 % (measured at the Irsching power plant: 60.75 %). Considering the emissions of pollutants, i.e. carbon monoxide, sulfur dioxide, nitrogen oxides and hydrocarbons, fuel cells have clearcut advantages over conventional power plants with their elaborate exhaust-gas purification treatment.

Road transportation represents another important anthropogenic source of CO_2 and other harmful emissions. It is responsible for about a third of the CO_2 emissions of the industrial nations. Here, electric power could be employed as a locally completely emission-free power train concept. However, one must not forget that the batteries used for electric-powered vehicles may simply transfer the CO_2 emissions to the power plant which recharges them, so long as the power is generated using conventional fossil fuels. Furthermore, the high weight and low storage capacity for electrical energy of the batteries (and thus the limited cruising range of the vehicles), as well as the long recharging times of available batteries, represent additional challenges to this technology (see the infobox "Storage Capacity of Various Energy Storage Devices" on p. 140).

Fuel cells, in contrast, make use of the energy stored chemically in hydrogen, or in fossil, biogenic and synthetic fuels, similarly to internal-combustion engines. Their high energy content makes it in principle possible for a vehicle with fuel-cell power to show driving performance competitive with that of conventional automobiles (see also the infobox "Storage Capacity").

For applications in transportation, fuel cells with a low operating temperature are particularly favored, especially the group of polymer-electrolytic membrane fuel cells (PEMFC). They generate sufficient power even at normal temperatures. Another advantage of a fuel-cell power train is its more favorable efficiency profile in the partial-load regime, for example in urban traffic. Conventional drive trains using internal-combustion engines develop their best operating characteristics only over long distances at higher speeds. Therefore, hybrid vehicles are under consideration, not least for economic reasons, where the fuel cells would

be as small as possible, corresponding to the limited requirements of urban traffic.

Very small fuel cells with power outputs in the range of a few watts have thus far not been able to establish themselves on the market. On the one hand, an effective alternative in the form of lithium-ion batteries has asserted itself. On the other hand, still unsolved problems, such as the use of fuel-cell cartridges in cabin luggage on aircraft, have prevented their widespread application. Thus far, only a few applications in the leisure and military sectors are promising for mass-produced micro-fuel cells. Here, mostly direct methanol fuel cells (DMFC) are in use; they are a modification of the PEMFC, in which liquid methanol is utilized directly at the anode. In spite of its advantage of simple fuel management, the DMFC has a weakness which so far has prevented its wide-scale introduction onto the market: Its low power density leads in the end to higher system and operating costs.

Fuel Cells for Road Vehicles

For vehicular applications, the demands on a fuel-cell system are very rigorous. Decisive factors compared to stationary applications are the desired minimal volume and weight for a given power output. Furthermore, a fuel-cell power plant must be adapted to short operating times (stop-and-go operation), which is the reason that only low-temperature fuel cells are suitable, due to their distinctive dynamics. They must be ready to deliver power within a few seconds and able to follow rapid changes in power demand during acceleration and braking of the vehicle.

Along with the technical hurdles, the cost factors are most stringent in the vehicular applications area. Fuel cells are moving here into a market where conventional power-train systems already exhibit very low costs of less than 100 €/kW. The costs of fuel-cell drives are currently well over 1000 €/kW, and even their considerable efficiency advantage cannot adequately compensate these

Fig. 1 *A low-floor bus from the EvoBus corporation, with a H₂ fuel-cell power train. Shown here is a fleet of buses which operate in the Hamburg urban region* (photo: Hysolution).

INTERNET

Basics and history of fuel cells
www.americanhistory.si.edu/fuelcells
en/wikipedia.org/wiki/Fuel_cell

Homepage of the organization Fuel Cells 2000
www.fuelcells.org

German fuel cell pages with animations
www.netinform.de/H2/Wegweiser/
Uebersicht2.aspx (click on 'English')

Research institutions
bit.ly/Smz8GB
www.zbt-duisburg.de/en

FIG. 2 | THE HISTORICAL DEVELOPMENT OF FUEL CELLS

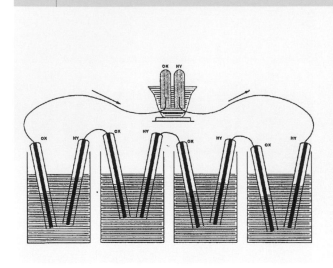 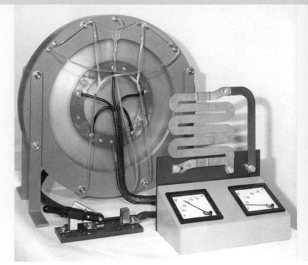

higher system costs. Ambitious programs to reduce costs and new technical developments however give hope that the goal of economic feasibility will soon be approached more closely.

The chief advantage of the "Polymer-Electrolyte-Membrane Fuel Cell" (PEMFC) compared to other low-temperature fuel cells is its straightforward cold-start behavior; however, the temperature should not drop below 0° C. At 20° C, they already deliver about half their nominal maximum power, and they reach their optimal operating temperature of 70–80° C in less than a minute. In addition, the polymer membrane which serves as electrolyte is mechanically robust and in general not sensitive to rapid temperature changes; thus, the PEMFC is predestined for non-continuous operating modes. The system can furthermore react to power-demand changes in the milli-second range, so long as the supply of fuel and atmospheric oxygen is maintained on this time scale.

A clear-cut weak point of the PEMFC consists of the fact that the membrane must contain water in the liquid phase. To prevent drying out of the membrane, the fuel must be moistened before entering the cell. This is accomplished by gas moisturizers as are shown in Figure 3. This limits their operating temperature range to between −20° C and the boiling point of water, 100° C. A further limitation is the fact that the PEMFC accepts practically only pure hydrogen as fuel. If one wants to use fossil, synthetic or biogenic fuels, these hydrocarbon-based substances must in most cases first be converted to hydrogen ("reformed") in a preceding chemical reaction. During reforming of these fuels using the steam reforming reaction to give hydrogen and carbon dioxide (CO_2), carbon monoxide (CO) is also produced in trace quantities (see also "The Production and Storage of Hydrogen" on p. 146).

Particularly at temperatures below 100° C, CO poisons the noble-metal catalysts in the electrodes of the PEMFC and thereby hampers the H_2 conversion reaction. CO is adsorbed preferentially to hydrogen on the catalyst and cannot readily be removed; the hydrogen is thus left unused. In order to prevent such a loss of power output, the reformed gas must be scrupulously purified of CO. This challenge was also one of the reasons why the automobile industry has in the meantime abandoned the idea of reforming fuels directly on board the vehicle.

Furthermore, one has to keep the heat balance of the PEMFC in mind. In order to avoid exceeding the maximum operating temperature of the cells, the heat of reaction and heat losses must be removed in a controlled manner. Although it has proven possible to integrate the elaborate cooling system in an elegant fashion into the overall unit, this problem remains an Achilles' heel for operation at high ambient temperatures.

PEMFC systems have in the meantime passed beyond the stage of prototype construction (Figure 3). The first broad-based projects for buses and bus fleets launched by well-known automobile manufacturers demonstrate the maturity of these systems. Parallel to these highly visible efforts, the hydrogen infrastructure required for widespread use is gradually being developed. The first hydrogen filling stations have been set up in several European cities. For this reason, it is not surprising that alliances between automobile manufacturers and hydrogen producers are mutually supporting the expansion of the hydrogen infrastructure. For example, the *Daimler* and *Linde* concerns are planning to construct 20 new hydrogen filling stations throughout Germany beginning in 2012. There is also an international plan to construct a network of stations along the Brenner Pass route from Munich to Verona.

However, the present price for hydrogen fuel is higher than for fossil fuels. This will not change in the near future, since hydrogen is currently produced with considerable effort by the reforming reaction mentioned above. It can

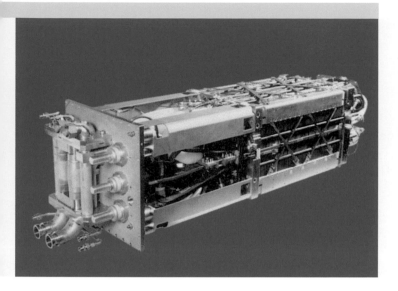

therefore not be cheaper than the raw material used as reactant. Furthermore, the stringent requirements for the safe handling of hydrogen add to its price. The costs of compression, storage and transport of the hydrogen are also not negligible.

The introduction of variable-availability energy conversion sources such as wind and solar power require storage of the converted energy for times when there is no wind or the sky is cloudy. Here lies a great opportunity for an alternative approach to hydrogen production (see also the chapter on "Hydrogen for Energy Storage"). During periods when surplus energy is being produced by the fluctuating renewable sources, it can be converted into chemical energy in the form of electrolysis hydrogen using suitable

HOW DO FUEL CELLS WORK?

Fuel cells are electrochemical energy converters. They convert the energy of a chemical reaction directly, i.e. without a thermo-mechanical intermediate step, into electrical energy. Normally, in an (exothermal) chemical reaction, the electric charges (electrons) are exchanged directly between the reacting atoms or molecules. So, for example, hydrogen reacts spontaneously with oxygen in the **detonating gas reaction**; the large amount of energy released by the oxidation of hydrogen is completely converted into thermal energy.

The trick in the fuel cell consists of not allowing the "fuel" to react directly with the atmospheric oxygen, but rather making it first give up electrons at the anode. Via the external circuit, the electrons flow through the power-consuming device and return to the cathode; there, they are taken up by the other reaction partner, typically oxygen from the air. In this way, the reaction can be carried out in a controlled manner and with a high yield of electric current.

The operating principle can also be considered to be a **reversal of electrolysis**. The construction of a fuel cell is very similar to that of a battery: It consists essentially of two electrodes

which can conduct electrons (anode and cathode) that are separated from each other by an electrolyte, which is an ionic conductor. The main **difference from a *battery*** consists of the fact that in the latter, the electrical energy is stored chemically in the electrodes, while in a fuel cell, the energy carrier is stored externally and the electrodes merely fulfil a catalytic function for its reaction. The structure of a fuel cell is shown schematically in the Figure using the example of a polymer-electrolyte-membrane fuel cell. The following chemical reactions take place in it:

Anode reaction:
$$2\,H_2 \rightarrow 4\,H_{ads}$$
$$4\,H_{ads} \rightarrow 4\,H^+ + 4\,e^-$$

Cathode reaction:
$$O_2 \rightarrow 2\,O_{ads}$$
$$2\,O_{ads} + 4\,H^+ + 4\,e^- \rightarrow 2\,H_2O$$

Overall reaction:
$$2\,H_2 + O \rightarrow 2\,H_2O$$

The simplest and preferred reactants are hydrogen as fuel and oxygen as oxidant. Purified

reformer gas (CO content < 100 ppm!) can also be used on the cathode side and air on the anode side. In technical systems, the fuel cells are connected together into "stacks", in order to obtain higher electric voltages and more power. Fuel-cell systems have a high efficiency, which in principle can be well above that of internal-combustion engines.

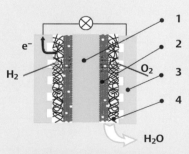

Structure of a PEMFC, schematic.
1: Polymer electrolyte,
2: Pt Catalyst,
3: Cell frame,
4: Current collector and gas diffusion layer

high-pressure electrolysis apparatus. In times of energy shortage, the storable hydrogen can again be used to generate electrical energy, either conventionally via thermal process cycles or directly with fuel cells. Large firms such as *Siemens AG* have recognised this weak point in a sustainable energy supply and have started major research programs on hydrogen production.

Fuel Cells for Stationary, Decentral Energy Production

In contrast to mobile applications, the requirements on fuel cells for stationary energy supply in a power plant can be more easily fulfilled, at first glance. In this case, we are not considering large central plants with an output power of more than 100 MW, since these can be more cost-effectively constructed using conventional technology, and they also operate with a high electrical efficiency.

Fuel cells offer advantages for the power range below 100 MW, which is typical of decentral power generation. In this range, the investment costs for conventional-technology plants relative to their output power increase rapidly, and at the same time their efficiency decreases. Here, system efficiencies of over 45 % at 100 kW output power and up to 70 % for fuel-cell plants with more than 1 MW are possible (using natural gas as fuel).

Such high efficiencies can be obtained when the waste heat from the fuel cells is available at a high temperature, so that a subsequent gas or steam turbine cycle can be used for additional power generation by the system. Alternatively, the heat from the fuel cells can be used for other thermal processes. Heat transport using adsorption and absorption

cooling systems for air conditioning of buildings is already being utilized (see the chapter on "Solar Air Conditioning"). In the least favorable case, the fuel cells can supply heat for warm water in a district heating network, along with electric power.

Operating temperatures above 600° C, which are typical of the so-called high-temperature fuel cells, initially have the advantage that a rapid gas reaction at the electrodes takes place even without an expensive noble-metal catalyst, and that the anode material can tolerate carbon monoxide. In principle, a variety of fuels can therefore be utilized. These include gasified coal and biomass, biogas and sewage gas, gas from coal mines and landfills, and synthetic or fossil hydrocarbons. At present, most installations are based on methane, either from natural gas as a fossil source, or from biogas and sewage gas as renewable sources.

Methane can be reformed within the fuel cell stack itself, making use of the heat produced in the cells. An externally heated reformer is thus not required, which saves on investment costs and to a great extent on operating costs. Furthermore, the requirements for cooling are reduced owing to the endothermal (heat consuming) nature of the reforming process. This leads to an improvement in efficiency in comparison to the PEMFC and PAFC systems, which operate at lower temperatures. On the other hand, higher operating temperatures have the disadvantage that the materials of the cells and the system components must be more corrosion-resistant and mechanically stable, which has thus far not been overcome. This makes them very expensive.

Stationary systems built by the *UTC* company (USA), with ca. 300 PAFC plants, are currently the most extensively installed, along with those of *Tognum-MTU* (Ottobrun, Germany) and *FCE* (USA), with around 100 MCFC plants. The SOFC and PEMFC technologies have been installed in only a few stationary systems to date.

The PEMFC

Decentral applications of the PEMFC for combined heat and power (CHP) production are limited by the low level of waste heat from this cell type. Since its waste heat is sufficient for low-temperature heating systems, PEMFCs could be utilized as household energy suppliers. Although PEMFC systems require a costly fuel pretreatment which reduces their overall efficiency, their suitability as household energy suppliers has been sufficiently documented in field trials.

All the well-known space-heating manufacturers have recognized this potential and are attempting to break into the market with mature technologies (cf. "Internet"). In these applications, PEMFC cells compete with SOFC systems. A generous market-introduction program in Germany [6] is supporting the installation of these systems. In a few cases, larger plants with over 50 kW of output power have been installed. Due to their high cost, no fuel cell applications have thus far achieved commercial success.

FIG. 3 | A FUEL CELL POWER TRAIN

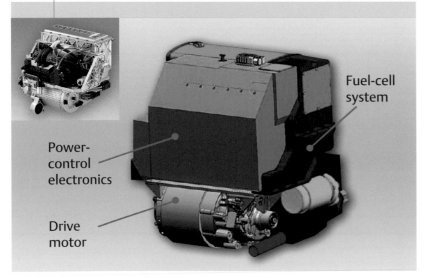

Fuel-cell system

Power-control electronics

Drive motor

A typical design concept for a PEMFC fuel-cell power train for road vehicles. The compact and integrated assembly of the individual system components has in particular led to a considerable leap in optimizing the system in recent years. Along with the fuel-cell unit (blue), the drive motor (gray) and the voltage converter from fuel-cell output voltage to on-board system voltage (orange) are combined into a compact arrangement [8] (graphics: Adam Opel AG).

The High-Temperature PEM

The high-temperature PEM (HT-PEM) is a hybrid between the PEM and the PAFC, which we treat in the following section. Here, the membrane serves only to fix the actual electrolyte, which contains phosphoric acid. This type of fuel cell operates at a higher temperature; it thus shows less sensitivity to typical catalyst poisons such as CO.

The Phosphoric Acid Fuel Cell (PAFC)

Owing to their higher operating temperatures of 200° C, the phosphoric-acid fuel cells can tolerate a considerably higher CO concentration of 1–2 % in the fuel gas. This means that an elaborate purification of the fuel gas downstream from the natural-gas reformer is unnecessary. Of all the types of fuel cells which are candidates for stationary energy generation, the PAFC is the most technically mature. Thus, *UTC* (USA/Japan) has already put a complete system (including reformer) with a power output of up to 400 kW on the market, and it has proved itself over years of operation. Unfortunately, here again the electrical efficiency is no more than 40 %, and thermal energy is output only in the temperature range around 90° C. Furthermore, the necessary investment costs are still far above those of conventional technologies.

The Molten Carbonate Fuel Cell (MCFC)

The MCFC takes its name from its electrolyte of alkali carbonates, which is used in molten form at the operating temperature of 650° C. The electric current is carried in the electrolyte by carbonate (CO_3^{2-}) ions. Maintaining the charge transport within the electrolyte thus requires CO_2 circulation between the fuel gas used and the air, which is usually achieved by using the CO_2-containing anode exhaust gas. The peripheral system components such as connecting tubes and heat exchangers require only a limited high-temperature serviceability; however, the extremely ag-

TYPES OF FUEL CELLS

Fuel cells can be classified as low-temperature and high-temperature cells. Low-temperature fuel cells operate in a range from room temperature up to about 120° C, while high-temperature fuel cells require an operating temperature between 600 and 1000° C. The phosphoric-acid fuel cell, with an operating temperature around 200° C, lies between these two rough groups.

Fuel cells are named for the type of electrolyte they use. There are alkaline fuel cells (AFC), the PEMFC (polymer-electrolyte membrane fuel cell), the molten carbonate fuel cell (MCFC), the phosphoric acid fuel cell (PAFC), and the solid oxide fuel cell (SOFC). Only the direct methanol fuel cell (DMFC) does not mention the electrolyte in its name, but instead indicates its ability to react methanol directly as fuel. Typically, it is based on a PEMFC. The different electrolytes determine to a large extent the characteristic properties of the corresponding fuel cells, such as their operating temperatures and conductivity mechanisms. Directly related to this is the choice of usable catalysts, requirements for the process gas, etc.

Low-temperature fuel cells require noble-metal catalysts in their electrodes (platinum or noble-metal alloys), in order to activate the electrochemical reaction. These catalysts are in general sensitive to CO poisoning at low temperatures.

The higher operating temperature of the PAFC, in contrast, allows it to tolerate CO concen-trations of 1–2 %, so that instead of pure hydrogen, so-called reformer gas can be directly input to the cells. The high operating temperatures of the MCFC and the SOFC make these cells insensitive towards CO and even allow the direct reaction of methane (natural gas).

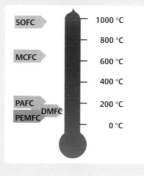

Typical operating temperatures of the different types of fuel cells and the properties which follow from them.

FIG. 4 | SOFC TUBE CELLS

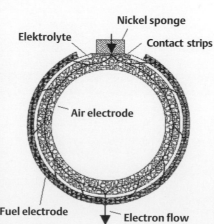

Left: This 100 kW SOFC plant (Siemens) has achieved a world-record operating life of nearly 40,000 hours (i.e. almost five years). For comparison: In the automotive sector, operating lifetimes of 5,000-10,000 hours are planned. Right: Cross section through a tube-shaped single cell.

Currently, hydrogen is produced on an industrial scale by steam reforming. In most cases, a fossil energy carrier such as methane is used as reactant for this endothermal, i.e. heat consuming reaction; it is chemically decomposed. Reforming can also be used as a preliminary step within a fuel-cell system. In the high-temperature cell types SOFC and MCFC, the operating temperature is already sufficient to supply the necessary process heat; this at the same time provides the required cooling of the cells. For low-temperature cells, in contrast, the necessary heat must be supplied externally, which reduces the system efficiency in the overall energy balance.

The resulting gas mixture is called reformer gas or reformate. Its main components in the industrial process are hydrogen, carbon dioxide, water vapor and about one percent of carbon monoxide. The technical alternatives to steam reforming are autothermal reforming and partial oxidation of fossil fuels. In industrial-scale processes, the fuel used is burned sub-stoichio-metrically, i.e. with reduced oxygen input, and provides the required process heat using suitable catalysts. The exothermal variant is called partial oxidation, while the energy-balanced sum of combustion and reforming is termed autothermal reforming. The latter has gained a certain importance in space heating applications. Operational differences, besides the time for preheating, are the different yields of H_2.

As byproduct of the reaction, CO is formed. The CO content at the output of the reformer is lowered by subsequent "shift" and purification steps from initially over one percent to values below 100 ppm (parts per million).

In principle, all organic hydrocarbons can be used for this process, as long as they contain no critical constituents (halogens, sulfur etc.). Technically established fuel-cell applications have employed among other fuels natural gas, sewage and landfill gas, biogas, mineshaft gas, methanol, ethanol, gasoline, LPG, diesel, or glycerine (from biodiesel production). Even the use of synthesis gas from coal is being investigated for SOFC hybrid power plants, especially in countries with a good supply of coal, for example the USA.

For H_2 storage, there are established methods using pressure vessels or cryogenic tanks. For special applications, metal hydride absorbers offer an alternative which permits the reservoir to operate almost without external pressure. In submarines powered by fuel cells, giant metal hydride storage cartridges are mounted on the outer hull walls.

Steam Reforming:

$$CH_4 + 2\,H_2O \rightleftharpoons CO_2 + 4\,H_2$$

Partial Oxidation:

$$CH_4 + O_2 \rightleftharpoons CO_2 + 2\,H_2$$

Autothermal Reforming:

$$CH_4 + x\,H_2O + (1\text{-}x/2)\,O_2 \rightleftharpoons CO_2 + (2+x)\,H_2$$

Various thermal processes for producing H_2. The reaction schemes are idealized.

TAB. | REFORMING METHOD

	Steam Ref.	Autothermal Ref.	Partial Oxid.
ΔH (kJ mol^{-1}) at 25 °C	165	0	− 318
Reformate (% H_2)	74	53	41

Comparison of possible reforming routes for methane: Energy balance and typical hydrogen concentrations.

---

gressive electrolyte causes corrosion of the cell and stack components.

MCFC installations with output power up to 2.8 MW have already been built; their electrical efficiencies were 47 %, and overall efficiencies (electrical and thermal) of over 85 % were measured. However, in over 20 years of trials of these installations, the goals for their operating lifetimes and costs have not yet been met.

The Solid-Oxide Fuel Cell (SOFC)

In the SOFC, a thin ceramic layer of yttrium-oxide-doped zirconium oxide (YSZ) serves as electrolyte, allowing the passage of oxygen anions at its operating temperature of 900 to 1000° C.

The power output of the systems currently being tested ranges from one kilowatt for household energy provision using natural gas up to a pressure-driven 200 kW unit. The latter can be hybridized with a gas turbine. The expected electrical efficiency is in the range of 70 % – thus far, 53 % has been measured.

Marketing of the SOFC technology is to be expected only in the longer term, in spite of the performance already achieved and the attractive perspectives for the future. In the past twenty years, it has not been possible to reduce its manufacturing costs sufficiently, nor to improve the materials durability to an acceptable level. This type of cell technology exhibits the greatest volatility in terms of the number of firms that are attempting to develop it.

Micro-Fuel Cells

For supplying low to very low power levels (1–100 W), fuel cells are also interesting, because they can make the high energy densities of hydrogen or methanol available to electric power consumers, e.g. for portable applications. If very long operating times are desired, then the combination of a fuel storage system and a fuel cell has advantages compared to a battery. Operation at room temperature and their rapid operational readiness would seem to recommend the PEMFC and DMFC cells as the only reasonable options for these applications.

Although initial demonstration experiments have been successful, nevertheless micro-fuel cells have not been able to displace the established Li-ion battery technology. In particular, the open question of the supply and disposal of the hydrogen cartridges has been an obstacle, and certification in certain environments has been only slowly forthcoming. For example, the use of pressurized containers in aircraft cabin luggage is not permitted. The manufacture of micro-fuel cells is closely related to the technology of semiconductor devices on wafers. Thus, arrays of microcells which can be connected in series or parallel can be produced.

Only the direct-methanol fuel cell (DMFC) has gained some support on the leisure market. Here, systems with a few 100 W output power could supply power to campers and small boats. The DMFC is also being used in military applications. Portable mini-systems for the individual power supply of communications and navigation devices are undergoing testing.

If there is a breakthrough in hydrogen storage or a quantum leap in the technical maturity and concept simplification of the DMFC, then a market for micro-fuel cells could be developed much more quickly than for vehicle power trains or for decentral energy supplies.

Outlook

Whether, and how soon, fuel-cell vehicles will take over the streets depends to a large extent – along with cost reductions – on the future price of the currently preferred fuel, hydrogen. Applications in stationary power generation are less complicated, where dynamics, cold starting and H_2 purity are less important. Here, however, the high specific costs have continued to delay a broad-based market entry both for decentralized large-scale energy production and for household energy supplies.

In the meantime, fuel cells are competing not only with the conventional technologies that they were supposed to supplant. More and more, other so-called "new technologies" are entering the race in the same direction, and they are to some extent more cost-efficient and technically mature; for example, space heating systems based on Stirling machines.

Summary

Fuel cells have reached a high level of technical development. The PEMFC has demonstrated its reliability for a series of new applications. The PAFC and the MCFC have already been field-tested in many plants of 100 kW and more output power; in the case of the SOFC, market entry has failed due to the lack of reliable materials and the resulting high costs. Therefore, in order for fuel cells to be economically competitive with established technologies of mobile and stationary energy conversion, a drastic cost reduction must be achieved, both for the fuel-cell stack itself and for the auxiliary systems required for their operation. In vehicle applications, the still open questions of fuel supply (infrastructure, H_2 production and H_2 storage) must also be addressed.

References

[1] Jules Verne, *L'Ile Mystérieuse* (*The Mysterious Island*), Hetzel, Strasbourg **1974** and **1875**.
[2] L. Carette, K. A. Friedrich, and U. Stimming, Fuel Cells **2001**, *1*, 5.
[3] J. Larminie and A. Dicks (eds.), *Fuel Cells Explained*, Wiley, New York **2000**.
[4] W. Vielstich, H.A. Gasteiger, and A. Lamm (eds.), *Handbook of Fuel Cells* (4 Vols.), Wiley, Chichester **2003**.
[5] history.nasa.gov/SP-4029/Apollo_13a_Summary.htm.
[6] www.now-gmbh.de.
[7] R. von Helmolt, Adam Opel AG, *Elektrische Mobilität mit Batterie- und Brennstoffzellenfahrzeugen*, 9th Kaiserslauter Forum, TU Kaiserslautern, November **2010**.
[8] R. von Helmolt, Adam Opel AG, GM and Opel Fuel Cell Vehicles – a Progress Report, Ulmer Electrochemical Talks **2010**, Ulm, June 2010.

About the Author

Joachim Hoffmann, born in 1958 in Mannheim, studied chemistry in Hamburg and obtained his doctorate there in 1989 in the field of synthetic organic electrochemistry. He was postdoctoral assistant in 1990 at the CNRS in Rennes, France. From 1991 to 2005 he carried out MCFC development at Tognum-MTU, from 2005–2008 SOFC development, and since 2009 PEMFC development, both at Siemens AG.

Address:

Dr. Joachim Hoffmann, Siemens AG, Industry Sector, Günther-Scharowsky-Straße 1, 91058 Erlangen, Germany. hoffmannjoachim@ siemens.com

Mobility and Sustainable Energy
Electric Automobiles

BY ANDREA VEZZINI

Electric automobiles are zero-emission vehicles as far as their operation is concerned, provided that their power comes from renewable sources. Modern electric drive trains are superior to internal-combustion engines, but their expensive batteries remain a bottleneck. The automotive industry is now investing considerable sums in their further development. – An insight into the current state of the technology.

Electric cars indeed operate emission-free when the power for charging their batteries comes from renewable sources (Figure 1). Even when their power is generated from fossil sources, they show a good CO_2 balance, since they utilize the electrical energy very efficiently. Our measurements at the Bern University of Applied Sciences in Biel, Switzerland on an all-electric city vehicle, the Mitsubishi i MiEV, which has been on the roads in Europe since 2010, clearly demonstrate this. They showed a reduction of about 60 % in CO_2 emissions compared to a standard model with an internal-combustion engine, provided that the electric power was supplied entirely from a modern gas and steam combined-cycle plant.

The automobile industry has in the meantime made it clear that they see the future in terms of all-electric drive trains, following an intermediate phase of hybrid technology (internal combustion plus electric power), and they are investing heavily. But the pioneers are having difficulties: Thus far, two models of Nissan-Renault and Chevrolet have

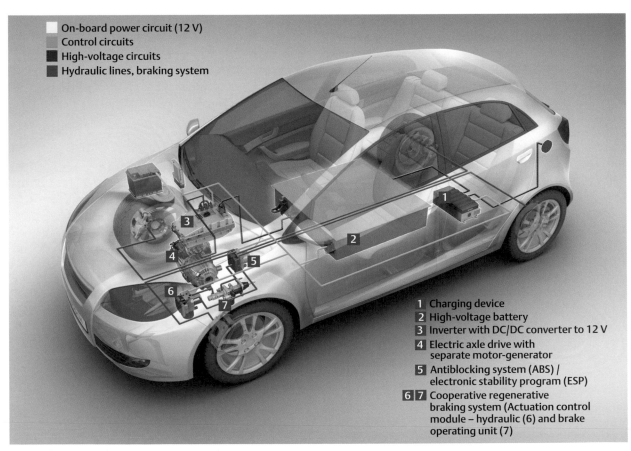

On-board power circuit (12 V)
Control circuits
High-voltage circuits
Hydraulic lines, braking system

1 Charging device
2 High-voltage battery
3 Inverter with DC/DC converter to 12 V
4 Electric axle drive with separate motor-generator
5 Antiblocking system (ABS) / electronic stability program (ESP)
6 7 Cooperative regenerative braking system (Actuation control module – hydraulic (6) and brake operating unit (7)

Fig. 1 *A "transparent" all-electric car: The large battery for drive-train power is located safely under the body between the axles. It and the drive motor operate within a high-voltage circuit at mains voltage level. The inverter (3) supplies the 12 V circuit including a conventional auto battery for peripheral devices (turn signals, window motors, instrument panel, multimedia etc.). When braking, the motor serves as a generator and recycles electrical energy (recuperation)* (graphics: Bosch).

Renewable Energy. Edited by R. Wengenmayr, Th. Bührke. Copyright © 2013 WILEY-VCH Verlag GmbH & Co. KGaA, Weinheim

fallen well behind the planned sales figures for 2011 [1]. A typical example, which represents this whole industrial sector, is the joint venture SB LiMotive, founded in 2008 by the auto-component manufacturer Bosch and the electronics firm Samsung and ended in September 2012; both companies now continuing its battery businesses alone. Starting with a development budget of 500 million dollars, they proposed to install a manufacturing capacity for lithium-ion batteries for 50,000 all-electric cars to be in place by 2013 [2].

Megacities with short-haul traffic and air pollution problems will be the first major markets for all-electric vehicles. The geographically small countries Israel and Denmark, for example, also want to electrify their road traffic, and they are cooperating with the Californian concern Better Place to set up an infrastructure of recharging stations and a battery leasing system [3]. Nevertheless, SB LiMotive takes a conservative tack and predicts that the world market for all-electric automobiles will take off in earnest only after 2020.

The batteries of future electric cars will be based on lithium-ion technology. According to the plans of SB LiMotive, for example, the first all-electric cars will have a cruising range of 200 km on a fully-charged battery. At an average daily driving distance of 90 km in Germany (30 km in Switzerland) for commuters, this should relieve the fears of potential buyers that they will be stranded underway. Additional obstacles to buyers are a lack of trust in the reliability of the batteries and the dashboard indicators for charge level and remaining cruising range, and the long recharging times, which are several hours. A serious problem at present is also the high cost of the large battery, which must be drastically reduced by the industry. This is emphasized for example among current electric vehicles by the Mitshubishi i MiEV: It costs two to three times more than comparable models with internal-combustion engines.

Experience with hybrid technology will pave the way in the coming years for all-electric drives. In hybrid vehicles, the power-control electronics, drive motors and an advanced battery technology are being tested to a degree which was previously unknown in the automobile industry. Electric drive trains are already superior to internal-combustion engines, since they deliver optimal power and torque over a much wider rpm range (Figure 2) [4]. Furthermore, they provide maximum torque to the wheels at stall. Thus, they require neither a clutch nor a large transmission. That reduces frictional losses and saves weight and space.

Modern electric motors are not only more compact than internal-combustion engines, they also have a considerably higher efficiency: They convert well over 90 % of the electric power into mechanical driving force, while the best diesel engines utilize only 35 % of the chemical energy stored in their fuel. To be sure, the same weight of fuel contains 50 to 100 times more energy than a present-day fully charged lithium-ion storage battery. Electric vehicles are

FIG. 2 | MOTOR CHARACTERISTICS

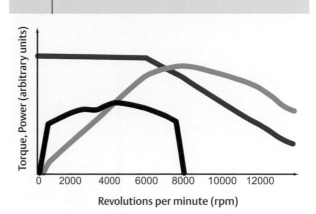

The torque (red curve) of an electric motor with 186 kW maximum power (Tesla Roadster), compared to that of a modern four-cylinder internal-combustion engine (ca. 120 kW maximum power, black curve); blue: the power curve of the electric motor (after [4]).

correspondingly heavy, even though their more efficient drive train requires less on-board energy.

An important step towards an all-electric vehicle will be the so-called plug-in hybrids. They are planned to travel over typical commuter distances using only electric power – including expressways. This already requires an electric drive train with comparable power to an internal-combustion engine. The drive battery must be correspondingly dimensioned. In contrast to self-sufficient hybrid vehicles, they will be recharged with electric energy directly from the power grid.

Drive-Motor Power and Battery Capacity

In designing the electrical drive components of a hybrid or all-electric vehicle, the drive power and the weight of the on-board battery must be optimized. One of the great advantages of an all-electric drive is the possibility of operating the drive motor for a short time at a multiple of its average maximum power. The average power can then be chosen to be considerably less than the peak-load power, making the electric drive train smaller, lighter and more economical.

We have simulated these requirements with software tools that we developed ourselves. If we insert the design electric drive into a power-consumption simulation, we obtain histograms for the energy used (Figure 3). For a quite realistic driving cycle developed at the ETH in Zurich (the Zurich commuter cycle), they show that the efficiency must be optimized for speeds under 50 km/h and for typical cross-country speeds of 100 km/h. The drive train must be optimized in particular for the partial-load regime. This also defines the region of average power, where the electric drive should have its best efficiency. For a city vehicle, this could be for example a drive power of 23 kW at partial load and 45 kW for peak load.

FIG. 3 | **ENERGY CONSUMPTION**

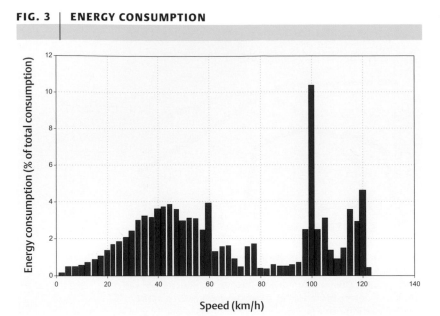

The distribution of the energy consumed vs. driving speeds for the Zurich commuter driving cycle (source: drivetek ag).

Once the components of the drive train have been defined, then the required battery capacity can be determined as a function of the maximum power needed and the cruising range desired. *Bosch* and *SB LiMotive*, for example, project a battery capacity of 35 kWh for a compact car with a 200 km cruising range and equipment and driving performance similar to those of a conventional automobile. This battery, however, including its technical peripheral devices, would weigh 350 kg. It is planned by *SB LiMotive* to reduce this to 250 kg; in addition, today's price of 500 €/kWh storage capacity is to be decreased to 350 € by 2015.

FIG. 4 | **BATTERY MANAGEMENT SYSTEM**

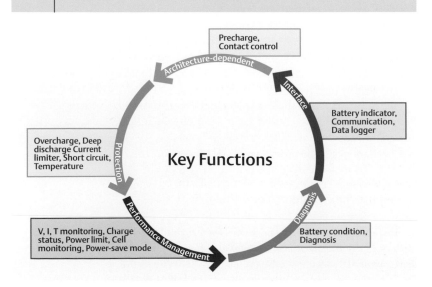

Functions of the battery management system (after [5]).

Then such a battery would still cost 12,000 € – it would thus still be a very major cost factor for the all-electric car.

The strategy of two other automobile manufacturers is also interesting: They announced serial-production plug-in hybrid models for 2010 and 2012. For the plug-in variant of the successful *Toyota* Prius, tests for the optimal design of the battery were carried out from 2009 to 2012. The driving behavior in daily usage of a 3rd generation Prius equipped with two tried-and-tested Ni-metal hydride (NiMH) battery packs of the previous series is being studied in cooperation with the French energy supplier *Electricité de France*. The NiMH battery weighs around 110 kg, while the future series production model using lithium-ion technology should weigh 55 kg for the same capacity, about 5 kWh. With electric-only driving, the cruising range lies between 15 and 20 km, and the test vehicles attain maximal speeds of 80 km/h. When their internal-combustion engines are running, they can drive faster.

General Motors is planning to extend the battery capacity of its Chevrolet Volt model from 16 to 16.5 kWh in 2013. Its current battery weighs in at 175 kg, corresponding to an energy density of 91 Wh/kg. High-energy cells today achieve up to 160 Wh/kg, but the weight of the whole unit, including all the control and safety electronics, reduces this to currently about 90 Wh/kg.

Battery Management System

These large high-power batteries require a battery management system with refined monitoring and protection functions for the battery (Figure 4 [5]). In normal operation, only the monitoring functions are active; their principal tasks are:

- *Voltage, current and temperature monitoring*: These primary data are used to derive other values. These include for example the internal electrical resistance, which gives information about the charge state and the general condition of the battery;

- *Charge equalization*: Since the characteristics of the battery cells are never precisely the same, and furthermore, the temperature may vary at different points within the battery pack, charging and discharging can load the cells differently. Over time, the capacity of the cells would drift apart, and the capacity of the weakest cell would limit the whole battery. Therefore, discharging resistances in parallel or actively-controlled recharging circuits provide for charge equalization among the cells.

- *Monitoring the charge state*: This plays a very important role in managing the battery and in the information display for the driver. While the battery management monitors the operation of the battery only within certain charge-state limits, a superordinate con-sumption computer calculates the remaining cruising range from the charge state.

The protection functions become active in case the battery is loaded outside its specified range, or in case of accident.

The battery must then be reliably disconnected from the rest of the high-voltage circuitry within the vehicle.

Lithium-ion Batteries

Storage batteries based on lithium-ion technology have many advantages. They for example have neither a memory effect like nickel-cadmium batteries, i.e. a loss of storage capacity after frequent partial discharge, nor an inertia effect as in nickel-metal hydride batteries. This 'lazy effect' causes the operating voltage to drop after recharging a battery which was previously not fully discharged.

In particular, the high energy-storage density of the lithium-ion batteries, already better than all their competitors, has been doubled within less than ten years, so that they dominate the market for portable devices such as laptops or mobile telephones. For hand-held tools, such as battery-operated screwdrivers, a similar development is rapidly taking place. In such tools, not only a high energy capacity (how long can I use the device?), but also a high power output (how strong is the device?) are required of the battery pack. Here, a variant of the lithium-ion technology, the lithium-iron phosphate battery, is increasingly dominating the market.

For hybrid and all-electric vehicles, the question arises as to whether these types of lithium batteries can simply be scaled up in their energy storage capacity and power output and used in the vehicles as energy storage systems. The requirements for the two types of vehicles are in fact quite different [8]. If the battery is expected to last for the life of the automobile, then in a hybrid vehicle, it must cope with up to a million charging and discharging cycles. In order to keep it from deteriorating, it is charged and discharged to only 5 to 20 % of its full capacity. This strategy would however make the battery of an all-electric auto much too heavy. During the life of the vehicle, it must withstand "only" 2,500 to 3,000 cycles, but it is discharged up to 80—100 % of its full capacity – and is thus considerably more stressed.

Furthermore, batteries for hybrid vehicles in particular must accept and deliver electrical energy quickly; they must be optimized for their power output. For all-electric autos, on the other hand, the battery must be able to store a large quantity of energy (how far can I drive?), i.e. it must have a high specific energy density.

Principle

Lithium is the most electropositive element in the Periodic Table. It yields the highest battery output voltages and thus also the highest specific energy densities, theoretically up to 1,000 Wh/kg. Lithium batteries function similarly to galvanic cells – to be sure, without a chemical reaction of the active materials. Instead, lithium ions are encased in the positive and negative electrodes as in a sponge. On charging (Figure 5, blue arrows), the positively-charged Li ions move from the cathode through the electrolyte to the spaces between the graphite layers (nC) of the anode, while the charging current delivers electrons through the external circuit. Within the anode, the Li ions form a so-called in-

FIG. 5 | **LITHIUM-ION BATTERIES**

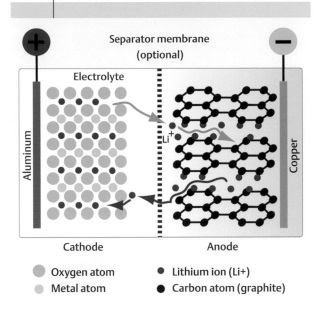

The structure of a lithium-ion storage battery. The separator membrane, which allows lithium ions to pass freely, is not necessary for the functioning of the battery, but is often inserted into high-power cells as protection against internal short circuits. This function can also be performed by the SEI layer on the graphite electrode (right).

tercalation compound with the graphite. In the case of a lithium battery with a lithium cobalt dioxide cathode, the following reaction formula applies:

$$LiCoO_2 + C \rightarrow Li_{1-x}CoO_2 + Li_xC. \qquad (1)$$

This reaction takes energy from the charging device and stores it in the battery. On discharging (Figure 5, red arrows), it is again released. The electrons in the external circuit can then perform useful work, e.g. mechanical work in an electric motor, before they flow back to the cathode. Within the cell, the Li ions drift back to the metal oxide:

$$Li_{1-x}CoO_2 + Li_xC \rightarrow Li_{1-x+dx}CoO_2 + Li_{x-dx}C. \qquad (2)$$

It is important for the proper functioning of the intercalation process that a protective covering layer be formed over the negative electrode, termed the 'solid exchange interface' (SEI) layer. It allows the small Li+ ions to pass through freely, but blocks the passage of the solvent molecules from the electrolyte. If this cover layer is incomplete, then solvent molecules are intercalated along with the Li+ ions, irreversibly destroying the graphite electrode.

Battery Types and Characteristics

For the usual lithium storage batteries, the active material of the negative electrode (the anode for the inner-cell circuit) is graphite. The positive electrode (cathode) usually contains lithium-metal oxides such as $LiCoO_2$, $LiNiO_2$, or

TAB. 1 | CATHODE MATERIALS

Cathode Material	Average Voltage (V)	Energy Density (Wh/kg)
$LiCoO_2$	3.7	110–190
$LiMnO_2$	4.0	110–120
$LiFePO_4$	3.3	95–140
Li_2FePO_4F	3.6	70–105
$LiNi_{1/3}Co_{1/3}Mn_{1/3}O_2$	3.7	95–130
$Li_4Ti_5O_{12}$	2.3	70– 80
Source: Leclanché SA.		

the spinel $LiMn_2O_4$, with a layered structure. Depending on the material, different energy densities and cell properites (high-power or high-energy battery) result. A summary of the typical cathode materials is given in Table 1.

For further considerations, I will group the usual cathode combinations based on cobalt, manganese, nickel or a combination of these elements under the collective term 'lithium-ion batteries'. Owing to the relative maturity of their technologies, they are currently at the focus of attention from the automotive industry. Batteries based on lithium titanate ($Li_4Ti_5O_{12}$) and lithium-iron phosphate ($LiFePO_4$) will be considered separately.

Compared to other battery systems, the lithium batteries exhibit a higher specific energy density for nearly all the cathode materials (Table 2). Still more important, however, is their often higher stability with respect to charge/discharge cycling. Here, the titanate technology is very promising, although it is not yet mature. The only lithium titanate battery currently available commercially is the Super Charge Ion battery (SCiB) of *Toshiba*. This battery is supposed to be chargeable to 90 % of its nominal capacity within 5 minutes. However, the SCiB has a relatively limited energy density of about 67 Wh/kg, and thus hardly more than Ni-MH batteries (Table 2).

Operating Lifetimes
Precisely with regard to cycling stability and operating lifetime, one finds differing results depending on the application. While the battery is usually very deeply discharged in all-electric or plug-in hybrid vehicles, in hybrids it is more frequently but less deeply discharged, as mentioned above [8]. Another important factor for the cycling stability of lithium-ion batteries is their operating temperature. Higher operating temperatures reduce the number of cycles that can be performed. Furthermore, the batteries lose some of their storage capacity even on the shelf – the higher the temperature, the greater this loss.

As a third factor, the charging and discharging power during operation limits the operating lifetime of the battery. At a constant charge/discharge rate of three times the nominal output current (3C loading), for example, the lifetime decreases depending on the particular technology by up to 40 % as compared to 1C loading. This is important, since the required drive power of an electric vehicle often exceeds the nominal battery power by a factor of 3 or 4. Nevertheless, this effect is usually negligible, since such peak power loads do not occur too often in practice, especially in city traffic.

Due to these complex interdependences, a model to predict battery lifetimes is difficult to set up. It would however be important for developing battery systems for all-electric and hybrid vehicles, since it determines significant commercial parameters such as the guarantee lifetime. In operation, also, the question of the "general health" of the battery is a central task for the battery management system. Furthermore, purchasers of used cars will want to know the state of the expensive battery.

Operating Parameters
The charging and discharging characteristics of the batteries at different temperatures and currents are particularly important during operation. They give a measure of the available quantity of energy – that is, how full the "tank" still is. However, only Li-ion batteries with cobalt cathodes have an approximately linear relationship between the charge removed and the battery voltage; this is no longer the case for cells with lithium-iron phosphate or titanate cathodes. Their charge state is registered by the newest battery management systems using an ampère-hour counter: It determines the charge removed from the battery by integrating the current over the time of operation.

TAB. 2 | BATTERY TYPES AND THEIR CHARACTERISTICS

	Lead-Acid Battery	NiCd	NiMH	Li ions	Titanate	Phosphate
Energy density (Wh/kg)	40	45–60	80	120–200	60–80	90–110
Cycles	200–300	1000–1500	300–500	500–1000	8000	2000
Temperature range (°C)	–20 – 60	–40 – 60	–20 – 60	–20 – 60	–30 – 80	–20 – 60
Self-discharge (%/month)	5	20	30	5	2	5
C-rate (power density)	Very high	Very high	High	Low	Very high	Very high
Nominal voltage (V)	2.0	1.2	1.2	3.6	2.3	3.3
Typical charging time	2.3–2.6 V: 20 h	C/10: 11 h	C/4: 5 h	4.2 V: 3 h	2.8 V: 1.5 h	4.0 V: 1.5 h
Output voltage (V)	1.7–1.8	1.0	1.0	3.0	1.8	2.0
C = nominal capacity of the battery. Source: *Leclanché SA*						

The amount of energy that can be extracted from the battery also depends strongly on its operating temperature. At low temperatures, the chemical reactions occur more slowly and the viscosity of the electrolyte increases. This increases the internal resistance of Li-ion batteries and lowers their available power output. The usual electrolytes can also freeze at temperatures below –25° C. Some manufacturers list the operating temperature range as 0 to 40° C; the optimal range is 18 to 25° C. There are however Li-ion storage batteries with special electrolytes which can operate down to –54° C. Thus, the battery management systems in all-electric and hybrid vehicles must include a thermal management program, which prevents all too drastic temperature fluctuations in the battery. In the winter, for example, the battery is heated while being charged from the power grid.

One of the advantages of Li-ion batteries is their simple charging procedures. Gassing out as with lead-acid batteries, which is difficult to control, does not occur. Therefore, the battery can simply be charged at constant current until it has reached its nominal voltage value; it is then further charged at constant voltage until e.g. the charging current has decreased to 1/20 of the nominal output current. Even with charging currents up to twice the nominal output current, around 85 % of the maximum charge can be transferred to the battery. Such a battery can thus be charged almost completely an about a half-hour. This is exploited for example by *Mitshubishi* in their electric car i MiEV. This can be done, to be sure, only with a special external charging device, since the usual mains connections can not deliver the required high charging currents and power levels.

In addition to this attractive possibility of rapid charging, the batteries exhibit a very high charging and discharging efficiency (Figure 6). The much poorer efficiency of lead-acid batteries is caused by losses during the gassing phase; in Ni-metal hydride batteries, the high internal resistance is in part responsible.

Furthermore, batteries for hybrid vehicles especially must accept and deliver electrical energy quickly; they must be optimized for their power output. For all-electric autos, on the other hand, the battery must be able to store a large quantity of energy (how far can I drive?), i.e. it must have a high specific energy density.

Safety

A frequent question regarding lithium-ion batteries is that of operating safety. Laptops that spontaneously catch fire and explode have badly shaken public confidence in this technology. There are varous causes for the failure of lithium-ion batteries, and its effects are likewise diverse. Often, metallic lithium, which reacts violently with water, is confused with lithium ions. The ions are considerably safer, assuming that the usual cautionary practices are observed. Among these are the avoidance of overcharging and of complete discharging. In both cases, metallic lithium precipitates onto the electrode surfaces. It forms needle crystals

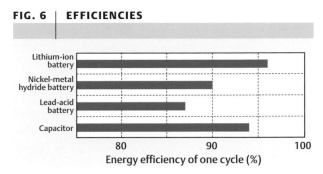

FIG. 6 | EFFICIENCIES

The charging and discharging efficiencies of various electrical energy-storage devices (after [6]).

there (dendrites), which can penetrate the optional separator and cause short circuiting and ignition of the lithium. A good battery management system will therefore keep a close watch on these limits and, if necessary, will disconnect the load from the battery.

The laptop fires were to be sure caused by metallic impurities from the production of the batteries, which gave rise to internal short circuits. This caused a temperature increase, starting a thermal runaway in which the cells released more and more energy. The cause of the open flames was mostly the flammable electrolytes (in particular the solvents they contained, such as ethylene carbonate, diethyl carbonate, etc.). Above 200° C, most cathode materials also decompose. Nickel-cobalt-aluminum ($LiNiCoAlO_2$) decomposes most readily, followed by cobalt oxide ($LiCoO_2$) and manganese oxide ($LiMnO_2$). Lithium-iron phosphate is the most stable. The graphite anode also produces heat and contributes to the thermal runaway; lithium titanate is less problematic.

The safety of lithium-ion batteries thus depends on their design and production quality. A good battery management cannot make a bad battery safer, but it can prevent damage to a good battery due to improper operation.

Conclusions

Recent progress with lithium-ion batteries in particular underlines the hope that all-electric cars will succeed on the market. This is especially true for short-haul traffic in urban centers. The positive experience gained with hybrid vehicles has encouraged automotive manufacturers to develop all-electric drives with a greater cruising range. The key to successful solutions lies in the consistent pursuance of two goals: Increasing the efficiency and reducing the empty weight of the vehicles. The integration of battery cells, control electronics and the heating and cooling peripheral devices into a single battery pack plays an important role.

It is also important that automobiles be designed in the future with all-electric drives in mind from the outset, rather than simply being refitted with with an electric drive train. In particular the peripheral devices and additional electrical loads on board, from the air conditioning to the window motors and power steering, should be designed especially for electric vehicles – then, the 12 volt on-board power circuitry including the lead battery could be dispensed with (Figure 1). Furthermore, consistent lightweight construction is necessary, in order to at least partially compensate for the weight of the batteries.

A question which is occasionally discussed in the news media will most certainly *not* be the cause of failure of electric cars as a mass-transport medium: a scarcity of lithium. The world's largest lithium producer, *Chemetall*, has estimated that there is a sufficient amount of readily accessible lithium on the earth [7], presuming that lithium from recycling of used batteries is also utilized.

Another important question is that of increased electric power consumption due to a growing number of electric cars. The answer is surprising, as the example of Germany shows. If the German Federal government achieves its goal of a million electric cars on German roads by the year 2020, then the power consumption in Germany will increase only by an estimated one-half of one percent.

Summary

All-electric automobiles are zero-emission vehicles in operation, so long as they are provided with electric power generated without CO_2 production. But even when they use power generated from fossil fuels, they have a good CO_2 balance. The reason: Electric motors have efficiencies of over 90 % (diesel engines: 35 %), and they can even dispense with a transmission. The bottleneck is still their expensive and heavy batteries, with their limited energy-storage capacities. Rapid progress in lithium-ion battery technology gives hope, however, that there will be a breakthrough within the coming decade. Large automotive batteries require a complex battery management in order to operate safely and with a long lifetime.

References and Links

[1] News from Autoguide.com: bit.ly/MhvHLK, and Dailytech.com: bit.ly/T4E2rh (August 2012).
[2] SB LiMotive, Press Release of 13th September 2011, bit.ly/MO2OvR.
[3] www.betterplace.com/global-progress.
[4] www.teslamotors.com.
[5] www.axeon.com (check "Technology").
[6] T. Markel, A. Simpson, Plug-In Hybrid Electric Vehicle Energy Storage System Design, **2006**, Preprint: www.nrel.gov/vehiclesandfuels/vsa/pdfs/39614.pdf.
[7] www.chemetalllithium.com (check "Resources & Recycling").

About the Author

Andrea Vezzini is Docent for Industrial Electronics at the Bern University of Applied Sciences, Technology and Computer Science Division, in Biel, Switzerland. He has carried out research for over ten years on the integration of lithium-ion batteries into mobile applications.

Address:
Prof. Dr. Andrea Vezzini, Labor für Industrieelektronik, Quellgasse 21, CH-2501 Biel, Switzerland. andrea.vezzini@bfh.ch

INTERNET

How lithium-ion batteries work (animation)
www.sblimotive.co.kr/en/products/basic_principle

Solar Air Conditioning
Cooling with the Heat of the Sun

BY ROLAND WENGENMAYR

When the Sun burns down mercilessly, it heats up buildings especially fast. But precisely then, it also delivers the most energy for powering large air-conditioning systems.

In order to keep cool, the Americans in particular consume enormous amounts of energy. The air-conditioning systems of the roughly one hundred million households in the USA eat up nearly five percent of the annual US electrical energy production. Worldwide, the market for air conditioning is growing at a startling pace. Along with it, the consumption of environmentally harmful fossil fuels is exploding. But there is a way out: Solar energy can also be used for cooling.

When the sun is shining mercilessly, it not only makes people suffer from the heat, it also provides a large amount of technically usable thermal energy. What would be more obvious than to use this energy directly to power the air conditioning? Such installations would offer an ideal solution for households and industrial buildings, whose cooling requirements are determined mainly by the local climate, that is the momentary insolation and the air temperature. They are also interesting for houses which have no connection to the power grid. This is true especially in Australia, since there, only ten percent of the land area is served by the electric power network. A high percentage of the households supply their needs by using a diesel generator. In these houses, cooling by utilizing solar energy would yield clear-cut reductions in diesel fuel consumption and CO_2 emissions.

In regions with a good infrastructure, also, an alternative to conventional air conditioning can pay off – even in cooler Central Europe. "At present, in Europe over 300 systems are installed. A growing number of them operate in the range of low cooling power up to 20 kW, since meanwhile suitable cooling installations for this range are available. Here, the research and development efforts of the past years are bearing fruit" summarizes Hans-Martin Henning of the *Fraunhofer Institute for Solar Energy Systems (ISE)* in Freiburg, Germany. The ISE researchers for example constructed a solar-assisted air conditioning system for the University Clinic in Freiburg in 1999; it cools a laboratory building.

All solar-powered air conditioning systems operate on the same basic principle: an evaporating liquid takes up heat and cools its surroundings; if the vapor is pumped away and liquefied elsewhere, it gives up its stored thermal energy again there. Using these two steps, heat can be transported out of a closed room. Refrigerators and conventional air conditioners also operate on this principle. In their case, a strong electric compressor densifies the vapor of the cooling medium, so that it becomes liquid and releases its thermal energy outside the space to be cooled.

A solar-powered air conditioner, in contrast, has to get along without an electric compressor – after all, solar collectors do not generate electric power; instead, they yield only heat. These systems cannot operate entirely without electric pumps to move the cooling medium, but the pumps can be designed to be much less powerful than a compressor. Thus, even a relatively small photovoltaic installation can provide the necessary electric power.

The function of the compressor is taken over by so-called absorbers: these bind the vapor of the cooling medium – it is usually water – and give off the heat content of the vapor. When heated, they release the water again and thereby cool their surroundings. There are a number of solid and liquid absorber materials; in everyday life, we often see them in the form of silica-gel beads, which keep moisture-sensitive products dry in their packages. In order to use the absorption effect for solar-assisted air conditioning, researchers and engineers have developed several technical concepts. Experts class them as systems with open or with closed cooling cycles.

Closed Systems

Systems with a closed cycle pump cold water through cooling tubes in the ceiling and walls of a building, like a reverse heating system. For systems of this type, *SK SonnenKlima* in Ahlen, Germany has developed a small refrigeration plant which, with a maximum of 16 kilowatts of cooling power, can cool for example one floor of a hotel.

This absorption refrigeration plant feeds a separate cooling-water circuit via a heat exchanger. To produce the cooling power, it evaporates water in a chamber at low pressure at 5° C. The absorber medium, a concentrated solution of the salt lithium bromide, then absorbs the water vapor, after which it is pumped into a second chamber at a higher pressure. There, the actual "motor" of the cooling system, namely the heat from the solar collector, again separates the water from the salt solution. A portion of the cooling power has to be diverted to condense the water vapor released from the salt, after which it passes again into the low-pressure chamber, completing the cycle. Closed-cycle refrigerators of this type require only moderate solar actuating temperatures between 55 and 100° C, so they can be operated with low-cost flat solar panel collectors.

Renewable Energy. Edited by R. Wengenmayr, Th. Bührke. Copyright © 2013 WILEY-VCH Verlag GmbH & Co. KGaA, Weinheim

FIG. 1 | SOLAR COOLING WITH A DOUBLE BOTTOM

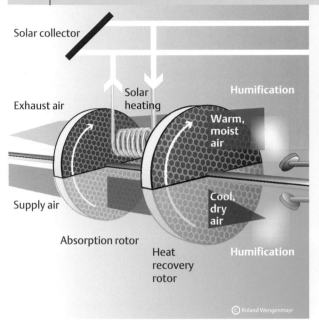

Solar collector

Exhaust air

Solar heating

Humification

Warm, moist air

Cool, dry air

Supply air

Absorption rotor

Heat recovery rotor

Humification

© Roland Wengenmayr

An open absorption-assisted cooling plant has an air supply duct (below, blue color) and an exhaust air duct (above, red). Two rotors turn through both ducts, the absorption rotor very slowly and the heat-recovery rotor around eighty times faster. Both have a honeycomb structure with many fine channels, in order to present the largest possible surface area to the air which is flowing past. The absorption rotor initially dehumidifies the incoming air, while the recovery rotor precools it. The dry air can now take up a large amount of water from the humidifier which follows. Since the water must evaporate in this process, it removes heat from the air: it cools and can be used for the air conditioning of the building. In order that the absorption rotor be continuously ready to take on water, the used exhaust air from the building is heated by a solar collector. Then it can take up additional moisture in flowing through the absorption rotor and thus carry off the water absorbed from the incoming air. The heat-recovery rotor and a second humidifier increase the cooling power of the plant.
(Graphics: Roland Wengenmayr.)

The Ahlen firm aims at a market price of under 10,000 € for their cooling plants; added to that is the investment for the solar collectors: All together, a small plant of this type is still more expensive than an electric compression cooling system of the same cooling power, which costs between 4,000 and 6,000 €. However, it requires only one-tenth as much electric power, and is therefore much cheaper to operate.

The *SK SonnenKlima GmbH* envisions a strong future market for small closed-cycle systems. In addition to the Mediterranean region, North America and Asia, also Arabian countries such as the United Emirates will be interesting markets for sales of the plants – as will Central Europe, as well.

Open Systems

Open systems draw in air, and not only cool the gas mixture, but also dehumidify it at the same time (Figure 1). They are suitable mainly for large central air-conditioning systems with a high air throughput.

For example, the factory of the furniture castor producer *H.C. Maier GmbH* in Althengstett, on the northern edge of the German Black Forest, has been air conditioned since 2000 using a solar-assisted open-cycle cooling plant. The researchers at the Fraunhofer institute, together with industry partners, installed a solar-assisted air conditioning system for the seminar room and cafeteria of the Chamber of Commerce and Industry of South Baden in Freiburg in 2001. These plants utilize an open absorption technique. Instead of using solar energy, plants of this type can also make use of industrial waste heat or other low-temperature heat sources such as surplus heat from a block heating and power plant for their actuation. A combination of solar heat with other heat sources is also possible. This flexibility opens a broad range of applications for solar-driven air conditioning even in Central Europe. Air conditioning using solar power is still exotic, but the number of these systems installed in Europe has nearly tripled over the past two years.

Summary

When the Sun is burning down mercilessly, it also delivers the most energy, which can be used to power large air-conditioning systems. At present, there are two basic methods of operating solar-assisted cooling plants. Closed systems work with a closed-cycle cooling-water circuit. They are principally suited for single floors or smaller buildings. For large buildings with a high throughput of air, open systems are more efficient: They cool and dehumidify the air supply directly. Both methods utilize the same principle: An evaporating liquid takes up heat and cools interior rooms or the water in a cooling circuit. The vapor is transported outside and re-condensed there, releasing its stored heat energy. The transport agents are absorption media – drying agents – which are freed of the water they have taken up by applying solar heat.

About the Author

Roland Wengenmayr is the editor of the German physics magazine "Physik in unserer Zeit" and is a science journalist.

Address:
Roland Wengenmayr, Physik in unserer Zeit, Konrad-Glatt-Str. 17, 65929 Frankfurt, Germany. Roland@roland-wengenmayr.de

Climate Engineering

A Super Climate in the Greenhouse

BY ROLAND WENGENMAYR

Fig. 1 *The Post Office Tower in Bonn, Germany, Is the first decentrally air conditioned high-rise building worldwide. The natural ventilation is driven by wind pressure and the chimney effect* (photo: Deutsche Post World Net).

Large buildings full of people and machines that are sources of heat are a challenge to air-conditioning engineers. This is especially true of alternative designs, which dispense with energy-consuming air conditioning and make the most of natural resources such as winds and the chimney effect.

Large buildings often present a challenge for providing a comfortable climate to the inhabitants. Their extensive glass façades turn them into greenhouses on sunny days. The heat output of human beings is also a factor which is not to be underestimated when many people occupy a building; furthermore, one must not forget the energy output from technical equipment such as computers. In conventional high-rise buildings, air conditioning machinery thus takes up about every 20th floor, and on all the other floors there are voluminous air ducts above the false ceilings. This air conditioning eats up money and energy for its construction and operation, and it causes health problems for many of the occupants of the buildings.

Architects have been looking for a way to solve these problems with intelligent architecture for some years. They want to make expert use of sources of heating and cooling from the environment, as well as various physical effects which can contribute to providing a basic air conditioning. Additional equipment can then manage the 'fine tuning' and compensate for peak loads; it can be much smaller and more energy economical. Such an alternative construction however represents an enormous technical challenge: It must provide comfortable conditions in hundreds of rooms reliably during every season of the year.

For this reason, architects work closely together with air conditioning and climate experts, for example with the firm *Transsolar Energietechnik GmbH* in Stuttgart, Germany and New York. This pioneering venture into sustainable air-conditioning technology was founded by the mechanical engineer Matthias Schuler and some of his research colleagues in 1992, coming from the University of Stuttgart. In the meantime, *Transsolar* can exhibit an impressive list of reference projects, ranging from the Mercedes Museum in

FIG. 2 | AIR CONDITIONING

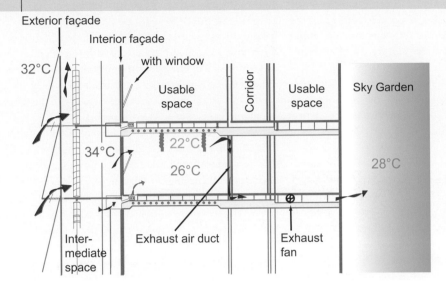

This cross-section through one floor of the Post Office Tower shows the path taken by the fresh air supply through the double façade into the offices (red arrows). The air first flows into the intermediate space between the inner and outer façades through the air shutters (left). From there, it enters the office spaces either through opened windows or through the subcorridor convectors. Via the exhaust ducts and the corridors it then flows on into the sky gardens, which transport it out of the tower using the chimney effect. During rare weather conditions, fans support the air flow. Water pipes in the concrete ceilings provide additional cooling for the rooms in summer (blue arrows); in winter, hot water is pumped through them for space heating (graphics: R. Wengenmayr).

Stuttgart, the main rail station in Strasbourg, to the international airport in Bangkok.

Every large building has to function like a single organism: Glass façade, roof, atria and stairwells, offices, conference rooms, cafeterias and the basement all become elements of an architectural air-conditioning system. Its task is to maintain the air in the building at a comfortable temperature and humidity and circulate it without creating unpleasant drafts. Unusual methods are employed, for example large channels in the ground which precool the entering air.

Beginning with the earliest planning phase, it must be guaranteed that all the components of the building will work together as planned. The engineers from Stuttgart employ elaborate computer models, which simulate the complete structure with all of its physical properties in detail. The software, developed by the firm, models the interior climate during the day and night, for every weather condition and in all seasons. The structural and physical properties of the windows, walls and ceilings and even the behavior of the people in the building are taken into account, insofar as they influence the interior climate. For example, the less effectively the architects make use of daylight, the more investment and power must be expended for illumination. Artificial lighting can with unsuitable planning become an important heat source within a building. The simulation of the interior illumination is therefore closely keyed to interior climate modeling.

In constructing very large buildings, the engineers are often breaking new technological ground, and computer simulations alone are not able to describe the situation correctly. In such cases, *Transsolar* builds real models of the planned building or its critical sections and tests them under varying weather conditions. If necessary, the engineers even construct a 1:1 mockup of a complete office with a

section of the glass façade and let it be "occupied" for several months by monitoring equipment.

The Post Office Tower in Bonn (Figure 1) gives an impressive example of what modern climate engineering can accomplish. The architects Murphy and Jahn in Chicago designed this new administration building for the German Post Office, a 160 meter high structure of 41 stories, and Helmut Jahn brought *Transsolar* on board. The result is summarized by *Transsolar*'s Thomas Lechner: "It is the first high-rise building with decentral venti-lation". This dry statement sums up a technical sensation: Instead of a central air-conditioning plant with its enormous air-supply and exhaust shafts, the rooms themselves, supplemented by many small air channels, secure the ventilation of the building.

Two physical forces keep the air for ventilation and basic air conditioning in motion: The wind that almost always blows around such a high building that stands alone, and the chimney effect, which causes warm air to rise in the interior of buildings.

In addition, they designed "activated" concrete ceilings: These contain thin water pipes, which carry cooling ground water from two wells below the building in summer, and hot water for heating in winter.

The architectural design accommodated this ventilation concept from the outset. The floor plan of the tower, which is supported by a pedestal structure, corresponds to two slightly shifted circular segments. Between them is a transitional area which contains very high spaces. These so-called sky gardens extend over nine stories each and form chimneys. The warm exhaust air from the offices rises up through them and then leaves the tower via exhaust vents on the sides. This gentle chimney effect is one of the two natural air-conditioning motors in the high-rise building (Figure 2).

The other natural driving force works at the air inlet side of the tower. There, the wind pushes fresh air into the building through some thousands of openings. In order to be able to use this pumping effect even under extreme weather conditions, the exterior façade consists of a double construction; the air first flows through ventilation shutters into the intermediate space between the façades (Figure 2). The north and south façades are separately regulated by a control system, which opens or closes the shutters depending on the wind velocity, the wind direction and the temperature.

The essential element for air conditioning is the many thousands of windows in the interior façade, which can be opened in order to adjust the interior climate. The occupants of the building can control the air conditions in each room individually. The windows are one of the openings for cross ventilation of the tower; if they are closed, then the air enters through a second opening (Figure 2): In the floor under each window is one of the 2000 "subcorridor convectors". These convectors are themselves small, individually adjustable air conditioning systems, which can heat or cool the air entering the rooms when the windows are closed. They have only a supplementary function. Ducts conduct the exhaust air into the neighboring corridor, where it passes through ventilation slits and finally into a sky garden.

Normally, the wind pressure and the chimney effect are sufficient to provide good ventilation within the whole building. On an average of thirty days each year, however, a lull in the wind combined with minimal temperature differences between the inside and outside of the building make additional ventilation necessary, and it is provided by fans in the exhaust air ducts.

In spite of this elaborate design, which was completely new, with the subcorridor convectors designed especially for this project, the contractors were able to construct the building at lower cost than with a conventional air conditioning system. The decisive point was that the decentral climate concept saves about 15 % in the interior volume of the building. This space is saved because the floors for air-conditioning machinery are not needed, and air ducts and false ceilings can be dispensed with. Of course, a portion of the cost savings went into the control systems.

The basic principles of the ventilation system are indeed surprisingly simple, but their realization in a high-rise building was demanding. The greatest problem is the enormous wind pressure. If the pressure difference between the windward and the leeward side were allowed to punch through into the interior of the building during an autumn storm, then many of the office doors would be pressed shut

and desks would be swept clear of papers. The climate engineers overcame this problem with a trick: They permit powerful air flows through the building in order to equalize the pressure differences. But these strong airflows lose their power in the intermediate space between the two façades, damped by the inlet shutters and the ventilation slits. This damping is so effective that the windows in the inner façade can be opened without causing problems even during a hurricane.

Transsolar had to demonstrate with certainty that their refined concept would be functional in practice. The air flow and temperature relationships in such a high building are much too complex to be able to trust computer models alone. The engineers tested physical models in a wind tunnel; the measurements were carried out by the Institute for Industrial Aerodynamics of the Technical University in Aachen. A decisive point was the demonstration that the air shutters would function as planned.

Since mid-2003, the Post Office Tower has been in use. Architectural psychologists from the University of Koblenz have investigated whether the roughly 2000 persons who occupy it every day are content with their working environment. The climate in the tower has been highly praised by all. In traditional Arabian houses, the chimney effect has been used for millennia to cool the interior; now it has also proved itself in a modern high-rise building.

Summary

Large modern buildings with glass facades, full of equipment and people, are a challenge to architectural climate engineers. This is particularly true of alternative concepts which dispense with enormous, energy-devouring air conditioning machinery and instead make use of natural resources. The Post Office Tower in Bonn (Germany) is the first high-rise building worldwide to utilize such a decentral air-conditioning concept. The chimney effect in its sky gardens and the pressure of the wind outside of the building provide passive air conditioning. Its concrete ceilings contain thin water pipes which carry cooling ground water from two wells below the building in summer, and hot water for heating in winter. Small, compact air conditioning units actively support the natural ventilation. The double façade even allows the office windows to be opened. Omitting the large-scale air conditioning machinery even allowed the 160 meter high building with its 41 floors to be built 15 % lower than a conventional structure with the same useful interior space.

INTERNET

Transsolar projects and further reading
www.transsolar.com

Cloud Spaces installation, Architecture Biennale 2010
blog.cloudscap.es/category/blog-tags/transsolar

About the Author

Roland Wengenmayr is the editor of the German physics magazine "Physik in unserer Zeit" and is a science journalist.

Address:
Roland Wengenmayr, Physik in unserer Zeit, Konrad-Glatt-Str. 17, 65929 Frankfurt, Germany. Roland@roland-wengenmayr.de

A Low-energy Residence with Biogas Heating
An Exceptional Sustainability Concept

BY CHRISTIAN MATT | MATTHIAS SCHULER

This example of a German therapy center including residential accommodation shows how an intelligent concept is capable of cleverly joining ecology and economy. Furthermore, the combined biofuel and cogeneration unit of the neighboring farmer makes even the heating largely autonomous.

The architectural office of Michel, Wolf and Partners (Stuttgart, Germany) has developed a complex of buildings in cooperation with *Transsolar Energietechnik GmbH* (Stuttgart) featuring an exceptional energy concept. It emerged from an old mansion in the Bergheim district to the northwest of Stuttgart. The plan was to renovate this villa so that it would accommodate a therapeutic institution for a group of just under 40 handicapped people. For accommodating the group residents and for other apartments, a new building was attached to the villa, built by the Deaconry Stetten. This religious institution wanted a building that would have low energy costs, use renewable energy sources, and still provide high living standards for its residents.

The new extension is a long, three-story, low-rise building whose generous windows provide a notion of transparency. These large windows fulfill two important tasks in our energy concept. On the one hand, they provide high-quality daylight for all the rooms of the new building, thus saving on artificial light and electrical energy. On the other hand, lots of sunlight enters the rooms in winter. This solar benefit additionally lowers energy consumption because heating can be reduced.

The fixed budget for the large low-energy home would provide a feasible solution only if we followed a holistic approach. From the start, all disciplines were included, i.e., structural work, electrical planning, as well as heating, ventilation, and sanitation. For example, the new building rests on a concrete channel instead of individual posts. Half of the channel serves as an earth duct, while the other half provides for technical infrastructure such as waste water, water for domestic use, and ventilation.

The earth duct beneath the building serves as a heat exchanger with the surrounding earth; it supplies the interior with fresh air. In winter, it pre-warms the air naturally before it is brought into the building via a ventilation system. The ventilation equipment provides sufficient fresh air supply to the residents at all times. At the same time, it is equipped with an efficient heat recovery system that greatly reduces the energy consumption compared to a window-ventilated building (Figure 1). Due to very effective thermal insulation and high-quality double-glazed windows in both the new building as well as in the old villa, the buildings lose very little heat in winter, and thus, their demand for heating energy drops.

In summer, thermal insulation and the outside sunshade keep the building cool. However, residents are required to play an active role. Our comfort concept asks them to open the windows themselves during night time in order to cool the massive concrete ceilings and walls through this night air flushing. In addition, the earth duct allows the ground to pre-cool the fresh air before it enters the building on hot summer days.

Fig. 1 *The new group home of the Deaconry Stetten with a sustainable energy concept (photo: © Lahoti & Schaugg, Esslingen-Stuttgart).*

The Heat Supply

Another distinctive feature is the building's heat supply: it has no heating unit. Instead, it is connected to the neighboring farm via a local heating pipeline. The farmer living there has installed a biogas system with a cogeneration plant. He runs an agricultural business with approximately 100 cows whose liquid manure delivers part of the 'fuel'. The remainder comes from organic waste of the sort that collects on a farm. From this, the biogas system produces methane gas (Figure 2). The downstream cogeneration plant generates sustainable electricity and heat from this methane gas.

However, the farmer was initially afraid of the investment costs, amounting to a good quarter of a million Euros, that would be necessary for such a system, in spite of his considerable personal contribution to its construction. Nevertheless, the Deaconry Stetten was able to convince him, since they as operators of the therapeutic institution guaranteed that they would buy the heat they needed from him on a long-term basis. This provides a reliable source of income to the farmer for amortization of his investment costs. It is supplemented with two additional guaranteed sources of income, thus making the project cost-effective. The farmer can feed the surplus electricity from the cogeneration unit into the grid and sell it; and since this electricity is 'renewable', it is additionally subsidized in Germany via the so-called Renewable Energy Sources Act (see also the introductory chapter in this book).

The biogas system features two further secondary benefits. First, it refines liquid manure to high-grade fertilizer that the farmer can distribute on his fields throughout the year. Second, the utilization of methane gas in the biogas system prevents this gas, generally produced by cows, from escaping into the atmosphere as is usually the case in agriculture. Methane is a particularly dangerous greenhouse gas, 20 to 30 times more potent than carbon dioxide.

However, the cogeneration plantt produces more heat than the Deaconry building needs for heating and hot water, particularly during the summer, even though the farmer also uses the heat to cover his private needs. This is due to the high insulation standards in the new residence and the renovated villa. Therefore, the farmer is looking for additional customers that could use the heat locally. In order to be able to guarantee the delivery of heat on winter days, the boiler in his farmhouse was supplemented with an oil reserve so that it can serve as a backup system.

CO$_2$ Balance

Apart from investment and operating costs, the CO$_2$ balance of the possible solutions was very important for finding the best concept. After the energy consumption for heating and hot water had been reduced by means of the

FIG. 2 | HEAT SUPPLY

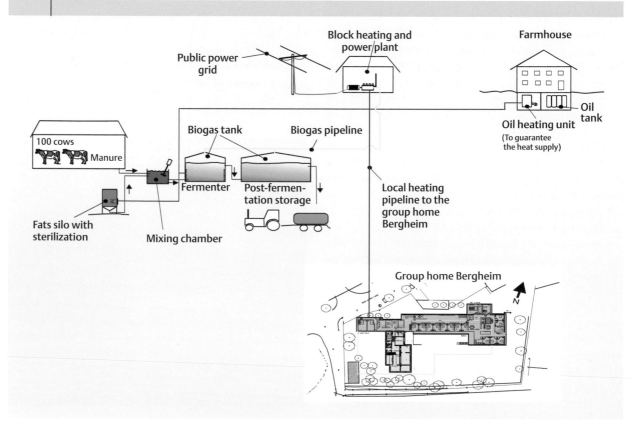

| *The local network heating concept with a biogas installation.*

FIG. 3 | **THE CO₂ BALANCE**

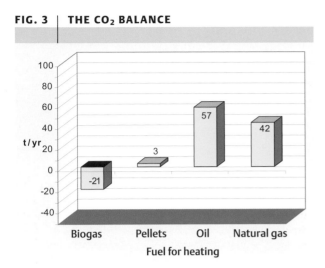

The annual greenhouse gas emission balances, expressed in equivalent tons of avoided CO_2 emissions, for various space-heating concepts. The 21-ton contribution shown at the left corresponds to the methane which is not released by the farm from the manure produced, due to the biogas installation.

low-energy concept, the heat required for the therapy center amounted to 161 megawatt hours per year (MWh/a), or, relative to the floor area, 55.5 kWh/m²/a. Here, even after renovating, the consumption of heating energy in the old building is twice as high as in the new building.

For assessing potential greenhouse gas emissions, all supply variants such as biogas, wood pellets, and also natural gas or oil were investigated in terms of their CO_2 emissions (Figure 3). Biogas was by far the best alternative; moreover, it even features a negative CO_2 potential, i.e., it releases less greenhouse gas as compared to the case that the therapy center had not been built at all. As mentioned, this startling result is due to the fact that without the biogas system, the farmer's livestock would release methane into the atmosphere, which now is not the case. This reduces methane emissions noticeably, and corresponds to savings of 21 tons of CO_2 per year. In the most unfavorable case of an oil heating unit, 57 tons of CO_2 per year would be emitted in addition.

The concept that was implemented thus now saves 78 tons of CO_2 per year, compared to a conventional building. This clearly shows how effectively such a sustainable neighborhood solution can protect the environment and simultaneously provide an autonomous energy supply. It is a good example of how problems are better solved in cooperation than alone.

The experience of three years of operation has demonstrated that the local heat supply from the biogas installation functions without problems. The heating costs are notably lower than with conventional systems, since they are independent of fossil energy sources (oil, electric power, natural gas).

Summary

The example of a therapy building for approximately 40 residents shows the advantages of an intelligent overall concept. The complex comprises an old, renovated villa and an energy-efficient new building. Large windows combined with good insulation use solar heat in winter and reduce electricity-consuming artificial lighting. Via an earth duct, the ventilation system beneath the new building pre-warms the fresh air in winter and cools it in summer. A neighboring farmer provides heating and hot water. He built a biogas system with a co-generation unit that converts the liquid manure of his cows and other organic wastes into sustainable electricity. The farmer feeds his surplus electricity into the power grid. This concept reduces the total greenhouse gas emissions of the building complex including the farmhouse tremendously. Thanks to subsidies, it has already become economically attractive in Germany.

About the Authors

Christian Matt, born in 1962, studied mechanical engineering at the Institute for Thermodynamics of the University of Stuttgart and wrote his Diplom (Masters) thesis on a solar-thermally operated ventilation system. Since 1996, he has been with Transsolar and has been a project leader there since 2002.

Matthias Schuler, born in 1958, studied mechanical engineering at the University of Stuttgart with a focus on technologies for rational energy use. In 1992, he founded the firm Transsolar in Stuttgart and has been its General Manager since then. He has taught at the Biberach University of Applied Sciences and the University of Stuttgart; After seven years as guest professor, since 2008 he has been Adjunct Professor of Experimental Technologies at the Graduate School of Design, Harvard University, Cambridge MA, USA.

Address:
Christian Matt, Transsolar Energietechnik GmbH, Curiestr. 2, 70563 Stuttgart, Germany.
matt@ transsolar.com
www.transsolar.com

Building Thermography Examined Closely

The Allure of Multicolored Images

BY MICHAEL VOLLMER | KLAUS-PETER MÖLLMANN

Building thermography using infrared cameras is becoming increasingly popular, since effective thermal insulation of buildings is in great demand. The gaudy false-colour images contain a great deal of information, but can easily be misinterpreted by those lacking in knowledge and experience. We give an introduction to the technology and its tricks on the basis of selected examples.

The technology of infrared imaging systems has developed rapidly since the mid-1990's. It has been driven by progress in microsystem technology, which has produced increasingly powerful detector arrays (sensors for infrared radiation) and read-out circuits. At present, a variety of camera systems for various applications are on the market [1]. They are used in many areas of technology and teaching [1-4]. For civil applications, IR cameras with megapixel detector arrays are already available.

In recent years, much less expensive cameras with a smaller number of pixels have come onto the market. Typical of these are systems in the price range under 5000 €, with 160 × 120 or even only 60 × 60 pixels, and reduced software options. This is not surprising, since contact-free temperature measurement is an excellent instrument for analyzing e.g. the thermal insulation of buildings (Figure 1).

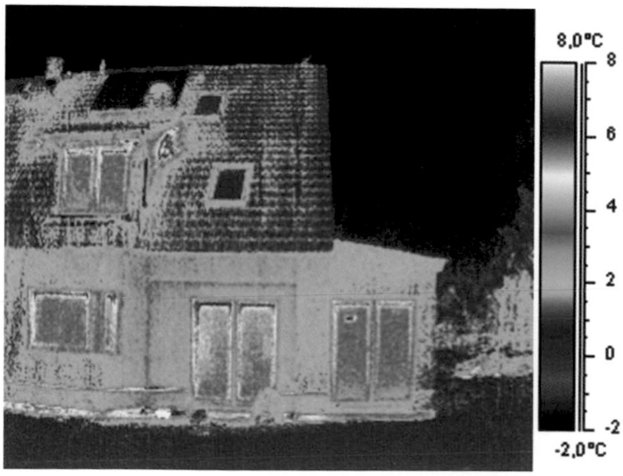

Fig. 1 ***Infrared images of buildings such as this one are frequently found in the media. The false-color representation is reasonable, but its interpretation is demanding. Untrained users often reach completely invalid conclusions.***

In this field, the word "infrared" is often replaced by "thermal", so that one usually speaks of thermal cameras and thermal images.

This basically positive development however creates some problems, in particular since the false-color images must be interpreted correctly. There are many traps lurking here, and we demonstrate some of them using examples. For this reason, several businesses and also the camera manufacturers offer training and certification courses on the fundamentals and applications of thermography [5]. To be sure, such courses cost as much as the less expensive cameras, so that some "services" dispense with the course and nevertheless offer thermographic analyses.

This is where the problem begins, as measurements carried out with an IR camera can be traced back to a variety of different sources of thermal radiation. Here, in addition to the object being investigated, the environment and even the camera itself play a role. In addition, there is a large number of possible error sources in recording and interpreting thermal images [1]. These can be avoided or corrected only by users who have a precise knowledge of the underlying thermal and radiation transport mechanisms. Untrained users, on the other hand, often simply produce colorful pictures.

The Goals of Building Thermography

In the course of the energy-efficiency discussion, buildings in the private as well as the industrial sector have become a focus of attention. Since they are usually heated using fossil fuels, better thermal insulation automatically results in a reduction of CO_2 emissions. Furthermore, fuel costs play an increasing role. This has resulted in political efforts to promote better building insulation (for German regulations, see e.g. [6,7]). One of the results is "energy passports", which document the energy consumption of buildings.

Fig. 2 *This infrared image reveals wooden half-timbering behind the stucco façade of this house. The different heat capacity and conductivity of wooden beams and masonry partitions gives a clear-cut contrast in the thermal image.*

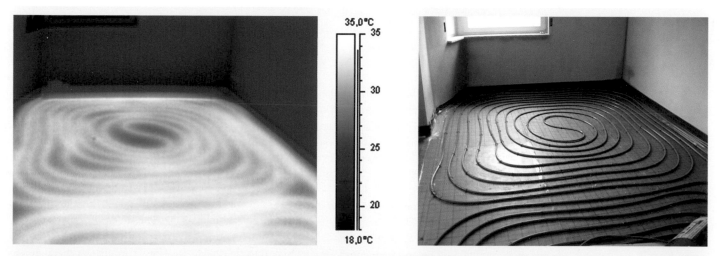

Fig. 3 *The infrared picture shows the piping of the floor heating system through the floor screed. The photo at the right, taken before applying the screed, demonstrates how precisely the pipes can be located from the IR image.*

Fig. 4 *The influence of radiative cooling from the environment on the surface temperatures of two neighboring buildings.*

The introduction of legal limits for heat losses has also necessarily led to the development of methods for testing the thermal properties and thus the insulation of buildings [8-11]. Thermography has come into use in this field since the 1990's for the following applications:

- Locating thermal bridges. This includes finding hidden structures such as half-timbering behind a stucco façade;
- Locating water intrusions and moisture;
- Identifying leakage from pipes, for example in floor heating systems;
- Quantitative determination of so-called heat-conductivity coefficients.

Methods

In order to be able to interpret thermographic images correctly, one must register a series of additional data before and during the measurements. A summary of the basic rules for structural thermography is given in Chapter 6 of reference [1]. Among these are general factors such as the preparation of the parts of the building to be investigated, but also geometric data. The latter include the absorption of the radiation by the atmosphere, the radiative influence of neighboring objects via the so-called view factor, and also the geometric resolution of the images.

Owing to their large masses, buildings or parts of buildings often have rather long thermal time constants, up to several hours. Therefore, weather conditions before and during the thermo-graphic investigations are important. Measurements on a dry morning before sunrise are optimal for quasistatic conditions, providing that it was cloudy and there was no precipitation during the night or the day before. This guarantees that the sun has not warmed up particular parts of the building by direct solar radiation. Furthermore, it ensures that there was no noticeable radiative cooling into the clear night sky. However, such conditions are not often available, so that as a rule, compromises must be made.

Exterior thermal images are often taken only to give an additional overview. Occasionally, for example with rear ventilated façades, they are not at all useful. Quantitative thermographic structural analyses are usually made from within the building, since many thermal signatures become visible only there.

It makes sense to take visible-light photos from the same perspective as the infrared images, in order to visualize the mapping in the test report. This report should take into account all the relevant ordinances and norms. It is usual to mark points, lines or surface areas in making a quantitative analysis of the images. This is done either directly in the camera or in a later analysis of the images. When all the parameters are correctly adjusted, these marked regions show either minimal, maximal or average temperatures. In order to capture the thermal signatures of small structures quantitatively with high precision, the optical system of the IR camera must provide the necessary geometric resolution [1].

Examples of Thermal Bridges

In the residence shown in Figure 1, one can clearly discern a reddish-white spot at the upper right edge of the dormer window. This is the thermal signature of an energy-effective heat leak, as can be demonstrated by comparison to other, similarly-constructed dormer windows on the same house (not visible in the picture). The cause is a lack of insulation. Interior thermography showed in the first instance that the temperature did not drop below the dew point when the outside temperature was around freezing; however, with still lower outside temperatures, this could happen. In addition, this heat leak caused such high additional heating costs that its repair would pay for itself within a few years. The rectangular dark areas, by the way, are reflections of the cold night sky from windows and a solar heating installation on the roof.

Hidden Structures made Visible

Thermography is by now a well-established method in monument preservation for making structures hidden behind walls visible. In Figure 2, the example of a half-timbered house is shown; the façade has been covered with stucco. The wooden beams and the filler material in the partitions between are clearly visible in the IR image, owing to their differing heat capacities and heat-transfer characteristics.

Thermography is also suited to the detection of other hidden building structures. Among these are the location of bricked-up and plastered-over windows, or finding piping in floor heating systems (Figure 3). Images of this type allow a view through the floor screed and floor coverings and can therefore serve to locate leaks.

View Factor and Thermal Time Constants

Figure 4 shows two neighboring houses, of which an outdoor thermal image was taken around midnight under a clear sky. The house in the background shows distinctly different wall temperatures from different areas, for example comparing the areas AR01 and AR03 in the image.

The cause in this case is not different insulation of the wall areas. A correct interpretation can be obtained only by taking into account the so-called view factor [1], which describes the radiation exchange between the walls and neighboring objects, the ground, and the night sky. The wall, which faces west, is not screened by any nearby objects, so that its radiation exchange with the cold sky is particularly strong. In technical terms, the night sky has a large view factor. As a result, the outer surface of this wall area (AR01) cools faster than the other wall area AR03, where the view factor of the sky is reduced by the neighboring house.

House walls with different structures also cool at different rates. These different thermal time constants are seen in Figure 4 in comparing the areas AR01 and AR05. The house in the foreground, due to the different composition of its walls, shows a notably higher wall temperature. The insulation of both walls is nevertheless comparable and sufficient; this however can be recognized only by analyzing indoor thermal images. The house in front has a rear ventilated façade. Due to the high heat capacity of the stonework façade, it has a much longer thermal time constant than the thin stucco layer mounted directly on the insulation of the other house. The image thus shows the remaining heat from solar warming during the previous day.

Consequences

As we have seen, a correct, detailed interpretation of thermographic images is a complex problem, since a number of different factors influence the results of a measurement. Therefore, the very low-priced thermal analyses being offered today, for example thermography of a house for only 100 €, must be regarded with some skepticism. "Analysis" is here often equivalent to simply taking IR images. The subsequent attempts at interpretation of the resulting colorful images without professional experience must then necessarily lead to incorrect conclusions.

Summary

Thermography of buildings using infrared cameras is becoming increasingly popular, as good thermal insulation of houses is growing in importance. The resulting garish false-color images contain useful information, but can lead to serious misinterpretation when examined by non-experts. A correct interpretation requires among other things information about the site, shadowing by nearby objects, the weather, and heating of surfaces exposed to the sunlight during previous days. As a rule, interior images are more meaningful. Along with evaluation of thermal insulation, thermography is also useful for visualizing hidden structures, such as covered half-timbering or heating pipes.

References

[1] M. Vollmer and K.-P. Möllmann, *Infrared Thermal Imaging – Fundamentals, Research and Applications*, Wiley, New York **2010**.

[2] D. Karstädt *et al.*, The Physics Teacher **2001**, *39*, 371.

[3] K.-P. Möllmann and M. Vollmer, European Journal of Physics **2007**, *28(3)*, 37.

[4] G.C. Holst, *Common Sense Approach to Thermal Imaging*, SPIE Optical Engineering Press, Washington, **2000**.

[5] See e.g. Infrared Training Center, www.infraredtraining.com and www.irtraining.eu.

[6] Deutsche Energieeinsparverordnung, Stand 29. April 2009. (In German), see: www.gesetze-im-internet.de/bundesrecht/enev_2007/gesamt.pdf.

[7] DIN 4108-7: Wärmeschutz und Energieeinsparung in Gebäuden, Teil 7, 2011-01, Deutsches Institut für Normung, Berlin **2011**.

[8] W. Feist and J. Schnieders, *Energy Efficiency – a Key to Sustainable Housing*, Eur. Phys. J.: Special Topics, **2009**, *176*, 141.

[9] A. Colantonio and S. Wood, *Detection of Moisture within Building Enclosures by Interior and Exterior Thermographic Inspections*, Inframation **2008**, Proceedings Vol. 9, 69.

[10] R. Madding, *Finding R-values of Stud-Frame Constructed Houses with IR Thermography*, Inframation **2008**, Proceedings Vol. 9, 261.

[11] Blower door technologies; see www.energyconservatory.com.

About the Authors

Michael Vollmer *studied physics at the University of Heidelberg and received his doctorate and Habilitation there. Since 1994, he has been professor of experimental physics at the Brandenburg University of Applied Sciences.*
Klaus-Peter Möllmann *studied physics in Berlin and received his doctorate and Habilitation there. Since 1994, he has been professor of experimental physics at the Brandenburg University of Applied Sciences.*

Address:
Prof. Dr. Michael Vollmer, Prof. Dr. Klaus-Peter Möllmann, Microsystem and Optical Technologies, Brandenburg University of Applied Sciences, Magdeburger Str. 50, 14770 Brandenburg, Germany.
vollmer@fh-brandenburg.de, moellmann@fh-brandenburg.de.

Subject Index